About Island Press

Since 1984, the nonprofit Island Press has been stimulating, shaping, and communicating the ideas that are essential for solving environmental problems worldwide. With more than 800 titles in print and some 40 new releases each year, we are the nation's leading publisher on environmental issues. We identify innovative thinkers and emerging trends in the environmental field. We work with world-renowned experts and authors to develop cross-disciplinary solutions to environmental challenges.

Island Press designs and implements coordinated book publication campaigns in order to communicate our critical messages in print, in person, and online using the latest technologies, programs, and the media. Our goal: to reach targeted audiences—scientists, policymakers, environmental advocates, the media, and concerned citizens—who can and will take action to protect the plants and animals that enrich our world, the ecosystems we need to survive, the water we drink, and the air we breathe.

Island Press gratefully acknowledges the support of its work by the Agua Fund, Inc., The Margaret A. Cargill Foundation, Betsy and Jesse Fink Foundation, The William and Flora Hewlett Foundation, The Kresge Foundation, The Forrest and Frances Lattner Foundation, The Andrew W. Mellon Foundation, The Curtis and Edith Munson Foundation, The Overbrook Foundation, The David and Lucile Packard Foundation, The Summit Foundation, Trust for Architectural Easements, The Winslow Foundation, and other generous donors.

The opinions expressed in this book are those of the author(s) and do not necessarily reflect the views of our donors.

ESCAPE FROM THE IVORY TOWER

ESCAPE FROM THE IVORY TOWER

A Guide to Making Your Science Matter

Nancy Baron

*With contributions by Liz Neeley
and the COMPASS team*

ISLANDPRESS

Washington | Covelo | London

ISLAND PRESS is a trademark of the Center for Resource Economics.

Library of Congress Cataloging-in-Publication Data

Baron, Nancy, 1964–
  Escaping the ivory tower : a guide to making your science matter / by Nancy Baron.
      p. cm.
  Includes bibliographical references and index.
  ISBN-13: 978-1-59726-663-5 (cloth : alk. paper)
  ISBN-10: 1-59726-663-9 (cloth : alk. paper)
  ISBN-13: 978-1-59726-664-2 (pbk. : alk. paper)
  ISBN-10: 1-59726-664-7 (pbk. : alk. paper) 1. Communication in science. 2. Science
news. 3. Scientists—Vocational guidance. I. Title.
  Q223.B37 2010
  501'.4—dc22

                                                                              2010001422

Printed on recycled, acid-free paper ✪

Manufactured in the United States of America
10 9 8 7 6 5 4 3 2 1

To my mother,
who would prefer that I write fiction.

Contents

Preface

The room was filled with twenty of America's brightest and most promising environmental scientists, including ecologists, resource economists, climatologists, and others. I was a magazine writer at the time, invited as the token journalist to help them see the world through a different lens. But the gathering quickly turned into a group counseling session.

Many of the scientists, it seemed, had a done-me-wrong story. One complained about being misquoted in an article. Another was blindsided by the negative reaction of his peers to the media coverage of his research. A third complained that she didn't get to review a news article that quoted her before it was published. The mood was one of fear and anxiety, borne of a lack of understanding of what journalists wanted or how to give it to them. I sympathized with the scientists who felt they had been misquoted or misunderstood. But I also understood from the journalist's perspective why it might have happened.

It was June 2000, and we were taking part in the second Aldo Leopold Leadership training session, established by leaders from the Ecological Society of America (ESA). Time and again, scientists had realized that they lacked the skills to navigate the rocky shoals of mass media exposure or to engage with policymakers. They recognized the problem but didn't know how to deal with it. The goal of the Leopold program was to give them training and to encourage them to lend their scientific expertise to inform public policy.

Jane Lubchenco was leading the charge on two fronts: She cofounded the Leopold program for environmental scientists and COMPASS, the Communication Partnership for Science and the Sea, an organization focused specifically on marine science. The original COMPASS partners were Sea-Web, Island Press, and the Monterey Bay Aquarium.

I was brought to the Leopold workshop, not just because I had been writing feature articles about environmental scientists but also because of my background. I had begun my career as a national park biologist in Canada working in both terrestrial and marine environments, which then morphed into science education and journalism. I began writing a "Field Notes" newspaper column for the *Vancouver Sun*, had a regular gig as an environmental commentator on the Global Television network, and wrote a series of feature articles about passionate environmental scientists. The stories about their research were a way for me to convey the urgency of issues such as the benefits of biodiversity or what's lost with the extinction of a species. My life's path had positioned me to straddle the worlds of science and journalism. But that path was about to take another turn. Although I didn't realize it at the time, that group counseling session with the Leopold fellows touched me deeply and set me on a decade-long mission to help environmental scientists communicate their science to the wider world.

As a practical matter, this meant giving up an adventurous life of magazine writing and moving from Vancouver, Canada, to California to become the first staff member at COMPASS, as its science outreach director. It was a tough decision. I loved writing features about environmental scientists and their work because I cared about their efforts and wanted others to care too. But my thinking went like this: by finding important stories in environmental science and letting other journalists in on them, I could have a multiplier effect in raising awareness that went far beyond what I could do as a lone journalist writing a single story at a time.

Also, the need was so great. Environmental scientists needed a new set of skills to help them figure out what to say, what not to say, and how to comport themselves outside the insular world of academia. They needed exposure to a range of journalists to demystify them and their worlds. I fashioned my role as that of a matchmaker, working to figure out more ways to bring scientists and journalists and scientists and policymakers together—and to coach scientists on how to get their message across to specific audiences.

This book is written from the trenches of that effort. It is the product of a decade with COMPASS, helping scientists communicate their research and developing media training workshops in the United States, Canada, and a half-dozen other countries. The Aldo Leopold communication sessions are one flavor of the workshops. Working with the Pew marine fellows was another. Other training and coaching sessions have been for faculty and students at universities, professional associations, foundations, and other organizations from around the world. Over the years, our team has grown and our COMPASS trainings have evolved to include new media and how to engage with policymakers, deliver legislative testimony, customize talks for nonscientific audiences, and better communicate across disciplines.

Although these activities can seem daunting to many, there is an exciting movement afoot. Increasingly I see nervous scientists transforming themselves into articulate, confident spokespeople for their disciplines. Today, environmental scientists including ecologists, economists, engineers, chemists, oceanographers, philosophers, epidemiologists, and social scientists are forming a new media- and policy-savvy culture of science. They are communicators, even leaders, informing and influencing policy. They know how to talk about their science in ways that can make people sit up, take notice, and care. Their delivery often appears easy and effortless. Yet, I know how hard they work to crystallize what is most important to say, and how best to say it. Some are reaching out to the public directly with new forms of media: developing public-friendly websites, writing blogs, and podcasting. Others become frequently quoted scientists in leading newspapers or contributors to opinion pages by writing op-ed essays. They have learned to move between the worlds of science, public policy, and the public debate with ease, confidence, and even grace.

I've witnessed this transformation over and over again. I regularly get e-mails from scientists telling me about their victories large and small as they engage with the media, lawmakers, or other government officials. They have important research, and they want to see it make a difference in the real world. We help them communicate it.

This book summarizes the content of training workshops my COMPASS colleagues and I have developed. Our goal is to help you get your message heard, appreciated, and perhaps acted upon. The book is designed to help you if you're just getting started or if you want to advance skills you've already acquired. And it provides a guide for scientists who want to teach their students these skills.

To be sure, there are other valuable, often more scholarly books on the art of science communication. Most of them fly at higher altitudes, discussing theories of science communication, and surveying the overall lay of the land.

This book zooms to ground level. It's a practical guide with specific directions on how to connect with journalists and policymakers and reach the broader public. It explores the ups and downs of deciding to engage with society and deals with the issues of advocacy, backlash, leadership, and—should you decide you want to—becoming an "agent of change."

This book would not have been possible without the experiences of many scientists who have participated in our workshops or through trial and error developed into brilliant communicators on their own. In the pages that follow they generously share their knowledge, their expertise, and their struggles. Most important, they share their lessons learned to help speed you on your way.

Acknowledgments

Many books are collaborative efforts, and this one is more than most. Its very genesis can be traced back to the formation of the Communication Partnership for Science and the Sea, or COMPASS.

I remain inspired by the vision of Jane Lubchenco, a marine scientist then at Oregon State University; Vikki Spruill, who at the time was president of SeaWeb; Chuck Savitt, the president of Island Press; and Julie Packard and Chris Harrold of the Monterey Bay Aquarium. Together with Jeannie Sedgewick and Mike Sutton, who were then part of the David and Lucile Packard Foundation, they conceived of COMPASS. The Packard Foundation has been a loyal funder of both COMPASS and the Aldo Leopold Leadership Program. I honor the foundation's ongoing support to environmental science and to expanding scientists' role in society.

This book is the product of working with COMPASS since its inception. When we set out on this journey, there was no charted course to follow. In moments of uncertainty, my mentor Vikki Spruill offered sage advice, her uncanny instincts, and encouragement.

I've been blessed by working with an amazing group of colleagues, scientists, and journalists. Many of their names, stories, and insights are in these pages. Many more are not.

Almost all the journalists in these pages have participated as trainers in my workshops. I sought them out and convinced them to become involved because they are among the brightest and best. Their wisdom and good humor infuse these pages. Gary Braasch, Jeff Burnside, Cory Dean, Juliet Eilperin, Tom Hayden, Jim Handman, Laura Helmuth, Chris Joyce, Natasha

Loder, David Malakoff, Michelle Nijhuis, Tim Radford, Andy Revkin, Erik Stokstad, Dawn Stover, Michael Todd, and Ken Weiss contributed for the love of the game. They share my conviction in the importance of scientific knowledge—and genuinely like hanging out with scientists and learning about their research.

Writing this book would not have been possible without the support of my COMPASS colleagues past and present. Brooke Simler and COMPASS principals Jane Lubchenco, Dawn Martin, Vikki Spruill, and Mike Sutton encouraged me and made it possible for me to tackle this book. I am indebted to Jessica Brown and Ashley Simons for their hard work, contributions, and camaraderie in the early days of COMPASS. Jessica developed many training tools and illustrated the concepts for the original workshops. Chad English provided the backbone of the policy chapters, which are now also COMPASS and Leopold training workshops. Emily Knight and John Meyer also helped significantly in the policy arena. Karen McLeod gave me moral support on book writing and provided scientific insights.

My teammates Liz Neeley and Matt Wright have been selflessly generous with their help. Always insightful, Matt Wright gave feedback on drafts and edited final versions of the chapters, deftly trimming my word whiskers. Without Liz Neeley I might still be working on the book. Her creativity and diligence led to many improvements big and small. Moreover, her companionship and enthusiasm made the late nights and weekends working on the book fun. I am also grateful to her for leading our charge into the world of social media and what it means for scientists.

Many friends and colleagues reviewed segments of the book and provided helpful feedback and content including Sandy Andelman, Dee Boersma, Dagni Bredesen, Pat Conrad, David Conover, Chris Costello, Chris Darimont, Paul Dayton, Andy Dobson, Simon Donner, Scott Doney, Jim Estes, Erica Fleishman, Steve Gaines, Ray Hilborn, Brian Helmuth, Jeff Hutchings, Elin Kelsey, Joanie Kleypas, Marty Krkošek, David Lodge, Meg Lowman, Pam Matson, Roz Naylor, Barry Noon, Bob Paine, Malin Pinsky, Jonathan Patz, Pete Peterson, Justina Ray, Bob Richmond, Enric Sala, Steve Schneider, Dave Secord, Sandy Shumway, Ben Santer, Reg Watson, Alison Watt, and Kate Wing. Thanks also to all the Leopold fellows and Pew marine fellows who shared their candid insights, not to mention wisdom and wine.

Ram Myers, Boris Worm, Jeremy Jackson, Daniel Pauly, Steve Palumbi,

Felicia Coleman, Larry Crowder, and Andy Rosenberg taught me by example and made me laugh through hell and high water.

A special thanks to Pam Matson, Pam Sturner, Margaret Krebs, Cynthia Robinson, Cynthia Barakatt, Polita Glynn, Margaret Bowman, Becky Goldburg, and Josh Reichert for giving me a chance to work with so many terrific scientific minds. I am honored to be part of their teams.

I am also thankful to the National Center of Ecological Analysis and Synthesis (NCEAS) in Santa Barbara for letting me work within its inspirational setting. What better place to be perched than at this veritable trap-line for scientists where I can pick the bright minds that come to work on issues of societal importance.

It's been a joy to work with Todd Baldwin, my editor at Island Press. Todd and Island Press President Chuck Savitt made this book and many others possible by bringing science out of the ivory tower and into the real world. Theirs is a labor of love.

Don Kennedy, who graciously agreed to write the foreword, has successfully straddled the fault lines between science, media, and policy during an extraordinary career. His is a model for any scientist who wants to play a larger role in society.

These names are not all inclusive. Many more people have made contributions that mattered. I regret I cannot name them all.

While this book has been enriched by so many, any errors that have crept in are mine alone.

What I have come to understand is, above all, the importance of building relationships and the power of team efforts. I am grateful to all who have helped me along the way—and for their desire to see environmental science contribute to society and sustain the natural world.

Finally, my deepest thanks to Ken Weiss, my husband, best friend, and life's coach, who shares my concern for the future and wants to make the world a better place by telling powerful stories from which we can learn.

Foreword: Donald Kennedy

At one time it was very hard to get young scientists to focus on how to construct a realistic, inviting narrative of their research and how they did it. Back in the day, when I regularly found myself a member of committees examining PhD candidates about their dissertations, I devised a question that I soon came to sense was unwelcome. It was approximately this:

> Imagine that you are waiting for an elevator with a friend on the ground floor of a nice hotel. Your friend is intelligent, curious, and well educated—but definitely not a scientist. You will have the duration of the nonstop ride to the fifteenth floor to explain what it is you did, what it means, and why it is important.

Some gave remarkably thoughtful and clear answers, leaving me hopeful. Some others, alas, appeared uncomfortable—and in a few cases almost resentful that they should be asked such a question. What, they seemed to be asking, does this have to do with my experiment or my thesis?

Well, the answer is "everything." As the editor in chief of *Science*, I looked for these same attributes in the review articles and the Perspectives pieces, in which scientist authors put new research findings into a context in which their significance can be understood. There was another challenge when my editorial colleagues were dealing with the contributed research papers we needed to review. The research we published naturally had to pass all of the tests of sound science, survive rigorous peer review, and represent a significant advance on previous knowledge. But we wanted the papers to be convincing and understandable not only to the practitioners of a single discipline or specialty. The paper must also be readable to the inhabitants of related fields—and, even more challenging, to an interdisciplinary audience to appreciate the context, relevance, and novelty of the work.

An important message to young scientists is this: your ability to make a compelling case for why your science matters is essential, too, if you care about whether public support of your kind of science is likely to continue. But there is something more: that is not the only, or even the best, reason for scientists to offer clear explanations and evaluation of what they do. The news we get from polls about public attitudes toward science is not good, and it's getting worse. In addition to the murky conflict between creation "science" and evolution, public skepticism about the science of climate change has grown steadily over the past decade, encouraged by an active denial industry with corporate support.

This state of dysfunction doesn't have to be permanent, and there will be more progress as more scientists take seriously their obligation to explain their subject effectively.

Nor are things improving with respect to science and the media. The disastrous decline of the print news business has probably been harder on science than on any other reporting beat. As things stand now, much of the useful news on breaking science events comes from blogs and informal sources of varying trustworthiness. As one commentator on this trend has pointed out, we are moving from the journalism of verification to the journalism of announcement.

Learning how to talk to the media in this new universe will be even more challenging—and more important now that the relative influence of the "name-brand" large metropolitan dailies and network television has declined to favor internet sources, talk radio, and cable TV. That will introduce an interesting shift. Senior scientists of my generation got used to talking to what was, after all, an elite cohort of media people. The new cohort of scientists will need to get comfortable with local reporters and citizen groups, however much they might prefer to be tapped for an interview on National Public Radio. If nonspecialists are to "get it," scientists must be better able to "give it."

Nancy Baron has been helping scientists talk to the media and to their public patrons for a decade, and *Escape from the Ivory Tower* records a rich history of the work she and her colleagues have accomplished. Two of my former students have been part of the enterprises Nancy has built, as part of COMPASS and the Aldo Leopold Leadership Program. Her extraordinary leadership has changed the lives of young scientists; here she has produced a book that will help both senior scientists and their students. The following pages provide a realistic, contemporary guide on how to deal with journalists, interviewers, the Rotary Club, and even congressional committees.

An important part of the story emerges from the Aldo Leopold Leadership Program, the continuing seminar first put together in 1998 by Jane Lubchenco (now the administrator of NOAA) and other past presidents of the Ecological Society of America. The idea, supported initially by the Packard Foundation, was to do some serious thinking about the needs of those who will find themselves laboring at the contentious convergence of science and public policy. It became an almost instant success. One might have thought of it initially as a program predominantly for younger scientists— those who might have been previously discouraged by their mentors for taking time away from the bench. On the contrary, a surprising number of more senior people came to see it as a professional opportunity. More recently, Nancy and her COMPASS teammates are making these skills more widely available through workshops for graduate students, and postdocs, and for conservation, NGO, and government scientists, too.

My personal experiences with this program, and having some of Nancy's associates show my students and me what they are doing, is wonderfully reassuring about the shape of the future. Much of what this younger generation of science communicators will accomplish must take place in a new era of transparency and will be shaped through the brand new institutions of the Information Age. Some of the media people they need to confront, and most of the members of the public, will appear on blogs and new internet sites. Scientific content will have to be searched out through webinars or in meetings conducted on Second Life. In short, researchers who want to engage will have to deal with brand-new outlets in which people are learning about science and deciding how to employ it in thinking about making public policy, skills that this book can help to develop.

That may sound rather trying to scientists of my generation—but there is no shortage of good news in this picture. The old idea that science and science policy should be left to the scientists is disappearing, to be replaced by a generation of scientists who understand that gaining public support is essential to their own futures and to their prospective role in policy. The senior undergraduates I teach at Stanford are entirely unfazed by this future. In a recent search for a relatively senior university position in science, a majority of the candidates judged to be finalists had been Leopold Fellows. So we are moving into an interesting and exciting but unfamiliar future. To those who would be its leaders, the advice in these pages provides essential guidance.

PART I

The Scientist Communicator

Chapter 1

INTRODUCTION

Science is more than just fascinating knowledge, it is also useful knowledge. I believe passionately that science should inform our decisions.

—Jane Lubchenco

Do you think your science should be useful? Would you like it to influence public policy or gain widespread recognition beyond your peers? Or perhaps even sway public opinion and help steer the course of history or human behavior? That's what this book is about: learning how to make your science matter, rather than getting buried in the dusty piles of scientific articles that collect in drifts on shelves and forgotten computer files.

This is a time of great challenges and opportunities for scientists and society. President Barack Obama famously promised to restore science "to its rightful place" after a long period of being sidelined and "to listen to what scientists have to say, even when it's inconvenient—especially when it's inconvenient" (Obama 2008). It is time for the very best scientists to engage.

Science is on the front lines as the U.S. Congress and state governments are paying increasing attention and debating climate change and other environmental and sustainability issues. The calls are mounting for scientists to talk to decision makers, provide testimony, answer journalists' questions, and help inform the public on issues of societal urgency. Yet there is a dearth of scientists who can deliver their information effectively and are willing to speak out.

On the heels of her confirmation as chief administrator of the National Oceanic and Atmospheric Administration (NOAA), Jane Lubchenco offered her perspective: "Decisions are going to take into account a number of things—values, politics, economics—but science should be at the table in a way that is understandable and relevant and credible and salient" (Witze 2009).

Lubchenco long ago became convinced of society's need for scientists to be more engaged. She sacrificed time from her research to assume leadership roles where she espoused the importance of communicating science. Her first call to arms came in a 1998 *Science* paper, "Entering the Century of the Environment: A New Social Contract for Science," where she argued that scientists need to be more forthcoming and share their research to benefit government managers, policymakers, and society at large. She urged her colleagues to invoke "the full power of the scientific enterprise in discovering new knowledge, in communicating existing and new understanding to the public and to policymakers, and in helping society move toward sustainability through a better understanding of the consequences of policy actions—or inaction" (Lubchenco 1998).

But Lubchenco didn't enter her field to change the world. Her story began in much the same way as yours did, most likely. You were probably driven, at least in the early days, by an intense desire to follow your curiosity. You dove into an all-consuming passion, pursuing answers to the questions that intrigued you—learning for learning's sake.

But then, perhaps you reached a second stage marked by a creeping awareness that what you are studying is changing—and likely not for the better. This is particularly true if you study the natural world. Environmental researchers, witnesses to nature, are often among the first to spot early signs when things are not as they should be.

At some point, your studies may have switched from understanding the natural aspects of a species, ecosystem, or physical phenomenon to investigating those changes that are cause for concern. Growing frustration and alarm may lead to the third stage—voicing your views as an expert.

"Two experiences motivated my decision to become engaged in science to inform policy decisions," says Barry Noon, an ecologist at Colorado State University who became embroiled in the spotted owl debate. "The first was a personal sense of loss over places that were important to me as a child—specifically, trout streams in Pennsylvania that I fished with my father. The second was a sense of anger over the distortion of science findings, stemming from research that I and others engaged in."

Yet moving beyond the safe, well-defined confines of research can be a difficult and even scary decision. Are you going to try to do something about the changes you are seeing? How do you reach beyond your research circles to communicate what you are observing to the wider world—why it matters, the potential risks, the possible solutions?

If you decide you want to inform those outside your research arena and help guide public discourse, you will need to learn a new set of skills. These include knowing exactly what you want to say, understanding your audience, and using common language to get your main points across clearly.

Ironically, just at the moment when science has more than ever to say about urgent issues, you must learn to navigate a world that is in rapid flux. The media is undergoing a revolution, and as a result, opportunities for scientists are simultaneously shrinking in some areas while expanding in others. Mainstream media outlets, especially newspapers, are in financial crisis, struggling to find new sources of income as subscriptions plummet and advertising goes to Craigslist and other places online. Still, mainstream media remain trusted sources of reliable information. At the same time, "new media" venues play an increasingly important role. They are instant, responsive, and effective at activating the power of the crowd. Often they riff off the mainstream media who do the original reporting. Questions still linger regarding the trustworthiness of some "new media," which is often heavy on opinion.

As the rules for what constitutes news are rewritten, the boundaries between old and new have blurred. Mainstream sources have begun to resemble "new" media by incorporating video, podcasting, and blogs, while online sources endeavor to earn credibility by instituting more formal editorial guidelines and processes. In light of such rapid and sweeping changes, only one thing is certain: the appetite for science news and information is alive and well, and if you can clearly and concisely articulate why your science matters, your message can transcend the medium.

Despite the stresses the news industry is enduring, local and national mainstream journalists still serve as gatekeepers. Talking to them remains a good way to reach other audiences. It identifies you to decision makers and politicians as an expert. A well-timed media story can break a logjam when government agencies are bogged down and unable—or unwilling—to act. Shining the media spotlight on an issue can force a resolution: whether or not to protect a species, to remove a dam, to allow roads to be built, or to challenge an environmental standard.

Journalists investigate important issues that are often out of sight and out of mind, like excessive harvesting of fish or forests, pollution, or habitat destruction, and bring them into focus for the public and policymakers. On the other hand, investing time in new media—whether firsthand by blogging or podcasting yourself, or talking to other preexisting outlets—can establish you as an authority who brings valuable viewpoints to the table. You can also build a following through social media, which raises your profile even further.

All media have their place in influencing society. Getting to know which media have the audiences you would like to reach can help guide you as you decide where to put your efforts. Ideally you can use both new and traditional forms to communicate most broadly.

Yet most research career paths don't prepare scientists to talk to the media, policymakers, or other audiences outside of academia. Some researchers have the horror stories to prove it.

Barry Gold, director of the marine conservation initiative at the Gordon and Betty Moore Foundation, learned the hard way about lack of preparation. When Gold was chief of the U.S. government's Grand Canyon Monitoring and Research Center, he led an effort to figure out how to restore portions of the degraded Colorado River ecosystem. "I got a call from the local press and I didn't take the time the night before to prepare for the interview," Gold reminisces. "At the end, the reporter asked me, 'What should be the headline?' I said, 'Gee, I don't know.' The next morning, there was the story on the front page of our local newspaper, with exactly the wrong headline and exactly the wrong message."

Alan Townsend, an ecologist and biogeochemist at the University of Colorado at Boulder, learned it's not *what* a scientist says, but *how* one says it. "I got a phone call from *Discover* magazine and in ten minutes I took the reporter from being really interested to mind-jarringly bored. He said 'thank you very much' and hung up. I never heard from him again."

Conveying important scientific insights to policymakers can be even more confounding, given their harried schedules, limited attention spans, and lack of background knowledge. Frank Davis, a professor at the Bren School of Environmental Science and Management at the University of California at Santa Barbara, bemoans a sense of "missing" when trying to communicate his science to policymakers and their staffs: "I talk to a lot to decision makers and I often get this feeling we are talking right past each other."

Scientists typically have one of two responses to poor results in dealings with the press and policymakers. One is sheer avoidance: skip that public hearing; ignore a request to meet with a policymaker; neglect to return a news reporter's phone call. Let's call this the ostrich approach. It usually doesn't serve anyone well. The second response, and the purpose of this book, is to rise to the challenge and learn the new set of skills.

As a scientist, you've learned much harder things because you're good at the most important element of success in this endeavor: preparation. If you prepare yourself for reaching out beyond the lab, field site, or office, you may join the ranks of top-of-their game scientists who excel at public discourse. You will most certainly increase the odds that you will competently convey an important message.

If this might seem like a stretch for you, I've witnessed this transformation over and over again in my longstanding role as the lead communications trainer for COMPASS and the Leopold Leadership trainings. I've seen nervous scientists who stumble in public discourse blossom into articulate, confident spokespeople for their disciplines.

A new breed of environmental scientist is emerging, including ecologists, economists, engineers, epidemiologists, chemists, philosophers, and social scientists who see the value in connecting their science to the world at large. They are communicators and leaders who inform and influence policy, and can talk about their science in ways that make people sit up, take notice, and care. Some are reaching out via new media, by blogging, doing podcasts, and putting a lot of effort into developing public-friendly websites. They are launching themselves into to a new orbit of engagement as communicators and leaders who move between the worlds of science and policy with ease, confidence, and even grace. And so can you.

The intention of this book is to offer you the motivation and, most important, the tools to get your information out, to be heard and understood. It can help you get started or advance your skills with practical advice on how to distill your core messages, talk to journalists and policymakers, prepare for an interview, write an op-ed, give testimony, prepare a "leave behind" for meetings with policymakers, promote a paper, anticipate and deal with backlash—in sum, to speak up for your science. These lessons will help you communicate what you know and contribute to making the world a better place.

Whether or not you feel prepared for it, society may come knocking at your door. Or you may be thinking about reaching out to the rest of the world because you care deeply and believe you can make a difference. The following chapters can help you learn the skills you need, increase your comfort level, and build your confidence so that you will be ready.

How to Use This Book

Scientists often assume that people will naturally understand *why* their science is important. But more often than not, they don't. The person who you are talking to (or perhaps at) is wondering, "Why are you telling me this?" When the answer is not forthcoming, people soon cease to listen or care. Watch them closely and you can see when their eyes begin to glaze over.

This book can help you get and keep their attention. It begins with a discussion of "culture clashes," comparing and contrasting the worlds of scientists with those of journalists and policymakers, and describes what these folks generally want from you.

The "how to" chapters offer simple and effective tools and strategies to engage your audience. With a little practice, researchers can communicate scientific information to policymakers in much the same way a journalist does: by asking (and answering), "What is the bottom line?" You can reach out through a wide range of ways, and this book gives you specific directions as to how. Our website gives you more examples and exercises to teach yourself or your students.

Embedded in each chapter are short case stories about other scientists' experiences and essays by journalists or policy experts that answer your most common questions, including:

- Why can't I read your news story ahead of time?
- How do I deal with scientific uncertainty?
- Why do journalists quote contrarians or others who don't know the science?
- What are the ups and downs of new media?
- Should I blog?
- Where do policymakers get their information?
- What do they want from me and how do I deliver it?

The final section offers advice on how to deal with backlash, as well as exploring the link between communications and leadership and how to become an "agent of change." This is grist to help you make some decisions, including what's the best fit for you and where you want to be. Throughout this book, you will see the icon 🖰 whenever we have additional context, examples, and related materials on our website www.EscapeFromTheIvoryTower .com.

If you feel that engaging with the wider world is part of your job—or are at least contemplating it—this book is for you. Hopefully it will also encourage scientists who are already publicly engaged and are looking for more tools and ideas to help them along. For graduate students and young scientists, who often need no convincing, it provides a how-to guide and exercises. For those who have participated in our communications workshops, it's a manual to help bring these ideas to your students and colleagues. And while this book is targeted at scientists whose work is related to the environment, it applies to anyone who is an expert and wants to make their science matter.

Chapter 2

THE DECISION TO SPEAK OUT

The saddest aspect of life today is that science gathers knowledge faster than society gathers wisdom.

—Isaac Asimov

Ransom "Ram" Myers was the last scientist one would ever expect to turn into a spokesperson for the world's fisheries. A self-described "math weenie," he loved "mining" and analyzing mountains of fisheries data. This PhD biologist found more than comfort in the numbers; he found the answers to solve problems. Data revealed truth.

For more than a decade, beginning in the late 1980s, he pored over fisheries statistics as a government scientist in one of Canada's most politicized agencies, the Department of Fisheries and Oceans (DFO). The numbers revealed a troubling trend: a centuries old staple fishery—the North Atlantic cod—was in trouble. Myers's investigations revealed that far fewer young cod were surviving than was previously believed, dashing future hopes for cod stocks facing chronic overexploitation. Between 1993 and 1997, Myers and his colleague Jeffrey Hutchings of Dalhousie University published twenty-eight peer-reviewed papers on the collapse of northern cod. But his conclusions were at odds with his superiors who believed that the numbers of upcoming young would be sufficient to stave off disaster. They ignored his recommendation that the department cut back catches to protect the fishery for the future.

"I tried to do an honest job of it," said Myers. The government had other plans. It tried to bury papers written by Myers, Hutchings, and DFO

11

fisheries scientist Alan Sinclair. It even went so far as to block them from presenting their findings that overfishing, not harp seals, was primarily to blame
for falling stocks at scientific meetings. Their science-backed conclusions
contradicted the government's party line. Myers broke ranks and went public,
which made a splash across the front page of Canada's national newspaper, the
Globe and Mail: "Overfishing, not seals, killed cod, buried fisheries report reveals" (Thorne 1997). Reprimanded by his superiors—including the minis-
ter of fisheries, who called the scientists, among other things, "prima don-
nas"—Myers had had enough. He left government work for the academic
freedom of a professorship of biology at Dalhousie University in 1997.

Myers's experience was far from unique. History contains case after case
of scientists who began their work in relative obscurity, only to be thrust into
the public eye for one reason or another. Reflecting on what was, perhaps,
the most difficult decision of his life, Myers acknowledged he wasn't really
prepared.

With a Darwinian beard and a preference for shorts and Birkenstocks, he
was never a dress-for-success type of polished speaker. He thought in numbers. His brain whirled at such high speed that his words tumbled over each
other, jamming at his mouth. When excited, he stuttered. Yet over time, he
learned to think through his material, chisel out his main points, prepare for
tough questions, and hone his message.

In 2003, he cowrote a paper with his graduate student Boris Worm about
how the global fishing fleet has removed 90 percent of sharks, billfish, tunas,
and other large fish (Myers and Worm 2003). It was to be published in *Nature*,
but Myers knew the world would not necessarily hear about it unless he explained what it meant and why it mattered. He left nothing to chance.

Like space, the ocean is remote from public experience. Myers wanted
the world to care about and pay attention to what was going on beneath the
surface. So he decided to put his all, albeit with a nonexistent budget, into
communicating his paper. "I want it to be like NASA," he said, referring to
how the agency learned to engage lay audiences in order to build support for
its costly and technical space program. With the help of COMPASS, he and
Worm prepared exhaustively for six weeks leading up to the paper's publication and continued communicating it long afterward, including responding
to critics. Today, many people still recall his passionate message: 90 percent of
the big fish are gone. "Speaking out effectively means doing an incredible
amount of work. By action, I do not mean simply mouthing off, but a long,

hard, sustained effort," reflected Myers. "Don't worry if people are mad at you, don't rub it in that they are wrong, simply produce so much evidence that your naysayers will eventually say that they believed it all along." Myers produced an amazing body of work of 160-plus primary literature articles that he authored or coauthored with his students before his death to cancer at age fifty-four in 2007. His passion for translating his mathematics into public awareness and actions lives on in his many students.

Now, drumming up public interest in your science may be the last thing on your mind. But there are multiple reasons to learn how to translate your work for broader audiences. It can alert the public and policymakers to issues of importance. The Myers and Worm paper generated waves of follow-up research and ultimately engaged broader public interest in sustainable fisheries and motivated chefs, retailers, and consumers to start making informed buying choices when it came to seafood. It also led to an invitation by Senator John McCain (R-AZ) to speak to the Senate Resources Committee because the media coverage was pervasive and people were talking about it. And so Myers and Worm flew from Dalhousie Nova Scotia to Washington, D.C., to present their findings.

U.S. Senate Committee on Commerce, Science & Transportation
Global Overfishing and International Fisheries Management Hearing
June 12, 2003

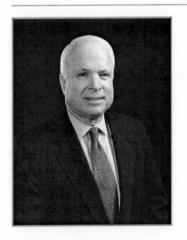

"Over the past month, there has been significant media coverage on global overfishing, which has helped to raise the nation's overall awareness of the condition of global fisheries. The message is our oceans are in danger and we need to take immediate action to protect them."

- John McCain (R-AZ)

Political figures pay attention to news trends, as John McCain demonstrated in 2003. Here he references the media coverage of Myers and Worm.

Learning to communicate is not only important to the outside world, but it can also enhance your stature and reputation among your peers and with your students. Everyone appreciates it when someone can sum up complicated matters in a succinct, cogent way. "Being able to communicate with nonscientific audiences greatly improves one's ability to communicate with scientific audiences as well—we are all human after all," says Ben Halpern of the University of California at Santa Barbara. "Learning to communicate is a critical life skill but it is not generally encouraged or taught within academic halls of science."

But learning to communicate effectively is really no different than learning a new methodology and applying it. Sarah Hobbie, a climate change scientist at the University of Minnesota, decided to pursue the Leopold Leadership training in order to learn the skills this book teaches. As she explains it, "I realized that I was avoiding situations in which I might have to engage the public, formally or informally, about my science. I lacked confidence." With some new insights and a bit of practice she proved to herself that she could do it:

> Communicating is a skill that can be approached methodically, critically evaluated, and broken down into steps that can be practiced. Sure, there are some people who seem born to interact with the media, but the rest of us aren't a lost cause. We can work to improve, as long as we have some guidance. You can make considerable progress in a surprisingly short time, if you apply yourself to learning and practicing a few basic skills.

The Changing Role of the Scientist as Communicator

The urgency of climate change, our need for energy alternatives, the degradation of the planet's life support systems, and the pressures of poverty and economic downturns mean that the knowledge and expertise of scientists of all stripes—natural, social, physical, and otherwise—are needed more than ever before.

Traditionally, scientists have viewed their role narrowly: ask questions that will advance the state of knowledge; then, go out and find those answers. Successful publication in a peer-reviewed journal was the end game, the final act

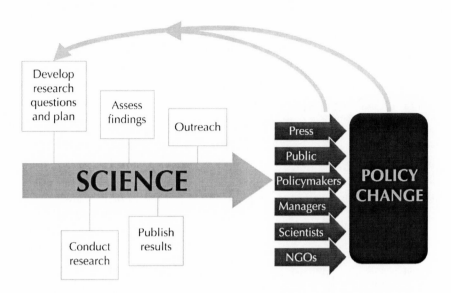

in a long and difficult process. Today, more scientists recognize the limits of this tradition and are moving beyond publication to communicate their results to the wider world.

Movers and shakers will not be dazzled by your data, your methodology, or the power of your writing in peer-reviewed journals. Politicians and decision makers, for the most part, don't read them. Journals are too complicated, too technical, and are not delivered to their doorsteps. These people need you to tell them what's going on and to answer their questions.

And they most often hear about you and your science through the news. Newly released studies in peer-reviewed journals often gain the most media attention because there is a consensus among members of the media that the peer-review process usually works and scientific journals are trustworthy. Since novel discoveries or advancements of knowledge qualify as news, if the media are alerted to a new paper and if they understand its relevance, they may cover it. But communicating your science in the media is *not* about over-reaching your results.

Scientists who become good communicators are sometimes accused of being enthralled with the limelight. Yet experienced scientists say that it is possible to achieve a balance, find a comfortable place between the conservative straightjacket of scientific decorum and the liberal splash of headlines, and ethically bring attention to various problems and scientific solutions.

Science *needs* help from scientists to find its proper place in society. Alan Leshner, CEO of the American Association for the Advancement of Science (AAAS), cautioned that "public skepticism and concern are increasingly directed at scientific issues that appear to conflict with core human values and religious beliefs or that pose conflicts with political or economic expediency." He sees a trend of weakening societal support for science and calls upon all scientists to go beyond their traditional roles. "It would be convenient to leave this task in the hands of a few representatives selected especially for their communication skills, but that won't work. Given the breadth of issues and the intensity of the effort required, we need as many ambassadors as we can muster" (Leshner 2007).

More and more key actors are of the view that a core aspect of doing science is communicating it. The National Science Foundation and the National Institutes of Health ask scientists to include a plan for outreach and public communication when they apply for grants. The research alone is not the end game. This is a trend also adopted by foundations such as the David and Lucile Packard Foundation, the Gordon and Betty Moore Foundation, and the Wilburforce Foundation, who fund programs that help scientists learn how to communicate to journalists and policymakers.

Universities, fellowship programs, and individual scientists are increasingly beginning to acknowledge the importance of communicating science. Many younger scientists are pursuing science careers with an eye toward how research and engineering can not only enhance our daily lives but also improve how we interact with our environment.

So if you want to share your wisdom with the wider world, you have to come out of your office, your lab, or your field station and take to the public podium. You will have to return reporters' calls, testify before committees of public leaders, and let your knowledge be heard. If you decide not to take these steps, you may be ceding the debate to those who know far less, or arrive with a self-serving agenda that shortchanges the best interests of society and the natural world.

How can the voices of scientists compete in arenas dominated by industrial lobbyists, influence peddlers, spin-doctors, and professional contrarians? First, you have a built-in advantage: scientists have deep knowledge on their side that comes from intensive and focused research. You are not simply voicing an opinion. Your insights and information should be considered in shaping public policy. Second, you have a natural ally, even though it may not seem that way—the media.

Science and the Fourth Estate

"Science and democracy have always been twins," writes Dennis Overbye in the *New York Times*, stating a widely held view among journalists (Overbye 2009). This is why reporters get frustrated when they encounter scientists who are reluctant to come forward. They believe that scientists have an ethical obligation to share their expertise and engage in debates about everything from human medicine to structural integrity of buildings to the preservation of rare and endangered species. Cornelia (Cory) Dean, a science writer and former science editor at *The New York Times*, believes that she does her part as a science writer to contribute to democracy, and scientists should do theirs. "So many of the issues we confront have to do with science, and if people aren't informed, they can't be good citizens," Dean says.

Journalists argue that the public has not only a need, but also a right, to know about what scientists do. When scientists receive public funds from

"I don't know why I don't care about the bottom of the ocean, but I don't."

federal agencies such as the National Science Foundation (NSF), the National Institutes of Health (NIH), or the Department of Energy (DOE), they have an obligation to share their findings with the public that helped finance the work. Journalists view themselves as representatives of the public. Their job is to go out and shine a spotlight on various issues for the benefit of their readers, viewers, and citizens at large.

Thomas Hayden, a former scientist turned science writer, says:

> One of my most emotional reasons for leaving science was that the knowledge I was contributing to was sequestered from the general public. I hope reporting and amplification helps build momentum and support for the work scientists do. If I can help you do what you do, that's good for the world at large.

For better or worse, journalists are conduits to other audiences. They are not your friends, they are not "on your side," but they are your connection to decision makers and the public. This is why this book places such a strong emphasis on communicating with journalists. It also teaches you how to talk directly to policymakers when opportunities arise—and when they don't, to realize the power of speaking indirectly with them through the media. Even in this changing world of online media, policymakers still track the mainstream media of record including the *New York Times*, the *Washington Post*, NPR, CNN, and increasingly many online and multiplatform media to help them set their agendas.

The "A" Word: Perspectives on Advocacy

You might believe your job as a scientist is done when you receive notice that your research is going to be published in a peer-reviewed journal. You might even think that you are undermining your credibility as a scientist to go any further.

Scientists agonize over issues related to advocacy: what is acceptable for scientists to do and what is not? Yet scientists aren't afraid of an open debate among their peers. It's the norm at conferences and seminars, or when reviewing and reacting to papers. Yet scientists cringe at the prospect of presenting what they know to the public or policymakers—for fear of being labeled an advocate. How, as a scientist, do you confront the reality that you also have values, interests, and a personality?

Pam Matson, dean of the School of Earth Sciences at Stanford and director of the Leopold Leadership program, believes, "We all need to be advocates for the *use* of science in decision making—and that's an easy place to be." Scientists, she says, must talk about their research in a way that's accessible to people who make public policy decisions. "If we start there, we are all advocates. But we are advocating for science—for the use of science and for the understanding of science. Then there is a broad spectrum from there."

So perhaps instead of getting stuck worrying about *whether* or not you should advocate, consider what constitutes *appropriate* advocacy.

The decision on how far to push your comfort zone is deeply personal. But you do need to think explicitly about this issue before you find yourself in a situation where you need to walk a line. Every time you are faced with an opportunity to provide information, including insights on the consequences and risks of any given action or inaction, you will have to make careful judgment calls.

Here, to provide grist for your mill, are some perspectives from scientists well known for their science as well as their views on communicating it. See what best fits you.

Roger Pielke: The Scientist as an Honest Broker of Policy Alternatives

In his book, *The Honest Broker: Making Sense of Science in Policy and Politics*, Roger Pielke, Jr., director of the Center for Science and Technology Policy Research at the University of Colorado, states: "Policy advocates will always seek selectively to use science in support of their agendas. . . . However, the scientific community itself need not view this process as the only mechanism for connecting research with decision-making" (Pielke 2007). He argues that scientists have a choice in how they engage with politics and society, and identifies the following four idealized roles:

- *The Pure Scientist*—focuses on research without any consideration of its practical use. He or she contributes to the peer-reviewed literature and has no direct interaction with decision makers.
- *The Science Arbiter*—wants to stay removed from policy, but will answer specific questions within the scope of his or her expertise. He or

she will interact with decision makers and the media, but only regarding questions that can be resolved by science.

- *The Honest Broker of Policy Alternatives*—participates in the decision-making process by clarifying and expanding the choices available to decision makers. He or she actively integrates scientific knowledge with stakeholder concerns and places information in the context of a wide range of policy options.

- *The Issue Advocate*—directly participates in the decision-making process by working to narrow the range of possible decisions and advance a specific political agenda.

Pielke argues that members of the scientific community fulfill each of these four roles, but that each choice is most appropriate under different scenarios of uncertainty, consensus, and political relevance.

If scientists want to maximize their ability to inform the views of decision makers, the honest broker's approach is usually most effective when high-stakes, value-laden debates are at play. It is particularly important for scientists to recognize and avoid claiming objectivity while using scientific authority as a tool to advance an agenda.

Stephen Schneider: The Citizen Scientist

"How can you enter the murky worlds of popularization, advocacy, and political negotiations—all of which are essential to get action . . . while at the same time preserving the soundness of the science?" asks Steve Schneider (Schneider 2003). Schneider, a professor of biological sciences at Stanford University and codirector of the Center for Environmental Science and Policy at the Freeman Sogli Institute, has spent decades tackling this issue in the politicized arena of climate change.

For him, the key is to clearly distinguish when you are talking about your science and when you are speaking as a citizen. "No one is exempt from prejudices and values but the people who know when they are bringing values and make their biases explicit are more likely to provide balanced assessments," Schneider says. "We must always admit where our expertise ends and where our personal value judgments begin, such as recommending specific policies. Policy choice is always a value judgment."

Schneider helps with Leopold Leadership trainings, role-playing a journalist catching a scientist off guard in an "ambush interview" after delivering congressional testimony behind closed doors. "What did you tell them in there?" he demands, thrusting a microphone into the scientist's face. Schneider wants to see if you can clearly deliver your main points. Did you stick strictly to the science or offer opinions? And did you clearly say which was which?

"It's up to us, as experts, to make the technical issues clearer through metaphor and simple language," he says. "We can't overstate the uncertainties on the one hand, nor neglect to mention dangerous or unpleasant possibilities on the other. Our job is to provide the context—and that often is in the eye of the beholder, so expect conflicts."

After years of experience, Schneider calls it like it he sees it. "Scientists who simplify to get heard will never succeed in pleasing everyone, especially not those colleagues who think scientists should stay out of the public arena whenever there is a call for simplification of the science. If we do avoid commenting entirely, then we abdicate the popularization of scientific issues to someone who is probably less knowledgeable or responsible. The bottom line is simply that staying out the fray is not taking the 'high ground'—it is just passing the buck" (Schneider 2008).

In a July 2008 article in *Nature*, "Getting It Across," policy expert Dave Goldston illustrates this philosophy of engagement. He describes how COMPASS helped scientists communicate the implications of ocean acidification to policymakers:

> Scientists spreading the word on acidification have been careful to maintain their credibility by being open about uncertainties and by drawing a line between science and policy. "The chemistry [of acidification] is highly certain," [Ken] Caldeira says. "But the biological consequences are highly uncertain." His goal, he says, is to make clear what the risks [of acidification] are. Even more importantly, COMPASS emphasizes that scientists should clearly distinguish between when they are describing science and when they are advocating what to do about the problem. Caldeira agrees: "I think that, as scientists, we have the ability and the right, if not the obligation, to speak as concerned and informed citizens. But it is useful to keep those roles separate. We have no particular priestly role where we have greater weight than anyone else" when it comes to policymaking. (Goldston 2008)

This distinction, as to whether you are speaking as a scientist or as a citizen, is useful. In any given situation, ask yourself: Which hat am I wearing—my scientist hat or my citizen hat? Then say so.

Daniel Pauly: The Scientist as Physician

Daniel Pauly, of the University of British Columbia Fisheries Centre, is among the world's most widely cited fisheries scientists. Arguably, part of the reason for this is because he is an outstanding communicator. His research has shaped science and policy by framing important concepts through evocative catch phrases: that fishermen are hunting at lower trophic levels by "fishing down marine food webs" (Pauly et al. 1998) and that most people have forgotten the long-term toll of overexploitation because of "shifting baselines" of knowledge (Pauly 1995). Pauly believes that scientists cannot fulfill their role by exclusively sorting through datasets. Instead, he says, scientists are morally obliged to communicate their work to the public. He points to the field of medicine as a guide to the question of whether scientists should be advocates for their work.

Environmental scientists should have their own Hippocratic oath, he says, "First do no harm." If scientists know that management or policy decisions will likely result in environmental harm, they have a professional and ethical obligation to stand up and be heard, and to speak up on behalf of "the patient"—in his case, the ocean and fisheries.

As physicians for the planet, "many environmental scientists have been too meek when managers, lobbyists, and politicians have challenged or contorted the results of their work," Pauly says.

> The main tool they have used to silence us is the notion that an engagement of the environment would compromise our scientific objectivity. Yet this argument is never invoked in medicine. Indeed passionate engagement for the patients against disease causing agents is not only the norm, but an essential element of doctors' professional ethics. (Pauly 2005)

Remaining silent is as unethical as the behavior of a physician who fails to speak up in the interest of saving a human life. "Scientists should be able to articulate the political implications of their science—this is not something

---

BOX 2.1
Two Reflections on Science and Advocacy

The following are personal points of view from a senior scientist, Steven (Steve) Gaines, and a young scientist, Chris Darimont. Both have been praised by some and criticized by others for advocating for their science.

Steve Gaines

The whole interface of science and policy gets messed up by this term of "advocacy." One of my guiding principles is to ask myself, how do we advocate for science, not for a position? Science gets marginalized when it should have a lot to say, and a lot of that is the fault of the scientists. You have to be a much more effective communicator to a broad range of audiences.

    The other issue is moving from the realm of what uncertainty means in science, versus what it means in the policy process. Scientists like to qualify everything by focusing on everything we don't know. I've learned to focus on what we do know. How do we go from the science we know and making the best possible decision, versus focusing on what we don't know, which allows others in the policy debates to dismiss the science?

    We know that some decisions are stupid and should be off the table. So science can put constraints on the possible decisions and point to those that make sense. By explaining that, science can rule out some options as being inconsistent with the scientific knowledge, even when there is uncertainty.

—Steve Gaines is director of the Bren School of Environmental Science and Management at the University of California Santa Barbara.

Chris Darimont

I don't agree with those who think their "objectivity" is threatened if they advocate. I am convinced that the peer-review system, though not perfect, provides a decent safeguard against our biases running amok. And if it does not, the great thing about science is that our papers are exposed to professional scrutiny at any time. So, I think it's a poor argument to not engage in media, for example, because it alters (the perception of) one's objectivity. My mentor, Paul Paquet, had some wise words about this subject: "Everyone's an advocate of a hypothesis, idea, or worldview and they express it in their decision-making, behavior, et cetera. Government biologists are often advocates of the status quo; industry biologists often are de facto advocates of resource extraction." So if I am viewed as an advocate of the maintenance of ecological and evolutionary process, that's fine by me.

BOX 2.1

Continued

I recall an undergraduate professor asking our wildlife management class who the next Jane Goodall would be. He was asking which one of us would stand up for the system we studied. And I was thinking, "Of course I would. It just makes good sense to feel that way."

Though others might not offer the following, I will. At least part of what motivates me to engage with media is an element of self-service. First, most people like attention, and on some level, I suppose I do too. Second, faculty job seekers are now increasingly rewarded for it. Not that this takes anything away from the drive we feel to be an agent of change, but its important that we are honest in acknowledging it. It certainly reinforces our other motivations.

—Chris Darimont is an NSERC postdoctoral fellow at the University of Santa Cruz, California, who studies ecological and evolutionary processes.

that should be reserved for policymakers who are often very far removed from the science itself." And, when scientists have something important to say, "it should then be said clearly, without equivocation" (Pauly 2005).

As Chris Darimont, an NSERC postdoctoral fellow at the University of Santa Cruz, says, speaking up for your science is not only an act of civic duty—it can also benefit you. Media stories featuring your work attract the attention of other scientists, and if the *New York Times* covers your science, for example, your citations get a bounce. Reaching out to the media can fulfill the outreach requirements that many funders demand in their grants. The increased visibility can help government and private donors recognize your work. They can make your institutional leaders happy and make your mother proud.

As some scientists are not afraid to admit, it can be a thrill to see your name in print, to see your face on the news, or to hear about your research as you drive to work listening to National Public Radio (NPR). Often, these bring on ripples of reaction from your friends, family, and folks you haven't heard from in years. Talking to people outside your field can also open your eyes to new dimensions of what is important and inform your research. While not all of your colleagues may pat you on the back, your students are likely to be supportive (Although this is not to say you won't be criticized, too). If

"I see by the current issue of 'Lab News,' Ridgeway, that you've been working for the last twenty years on the same problem I've been working on for the last twenty years."

your institution is not there yet, it's only a matter of time, and you can help lead the way. But the most important reason to speak out for your science is because society needs you and so does the natural world.

The Bottom Line

This book is not about spin or efforts to manipulate. You will never be able to control the media or policymakers. Honesty is the only sustainable policy. To be sure, things can go awry and sometimes do. But you can dramatically increase your odds of success. No matter how strong or weak you view yourself as a communicator, you can progress toward excellence and effectiveness. Just like playing a musical instrument or writing haiku, it can be a lifelong undertaking to up one's game.

PART II

A Clash of Cultures

Chapter 3

WHAT YOU NEED TO KNOW
ABOUT JOURNALISTS

Put it before them briefly so they will read it, clearly so they
will appreciate it, picturesquely so they will remember it,
and above all, accurately so they will be guided by its light.

—Joseph Pulitzer

Scientists and journalists may seem like they come from different planets, but
in many ways they are more alike than different—at least in terms of character. Scientists and journalists both love to ask questions and discover new
things. Both are analytical and skeptical. They
are typically highly competitive, driven person-
alities who strive to be first—whether publishing a new discovery or breaking a news story.

But when it comes to presenting their findings, scientists and journalists are utterly at odds.

Journalists literally want to know your bottom line—first. To talk with them, you must
turn what you normally do on its head and begin with the conclusion.

Since the beginning of your education as a
scientist, this formula has been drummed into
you: title, abstract, background, methods, results,

© Hadi Farahani

discussion, and, finally, the conclusion. In other words, you present your work
in exactly the opposite order that journalists want to hear it. These habits are

29

SCIENTISTS   JOURNALISTS

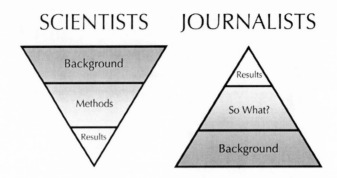

hard to break, and most scientists don't realize that they need to completely reconfigure their approach.

You must also make what can be a nerve-wracking effort to present information in ways that are simple, yet not simplistic. This generally takes some effort at translation, bearing in mind the interests of a nonscientific audience.

These cultural differences can bring the worlds of scientists and journalists into collision. To diffuse this conflict, it is helpful to take a look at where some of these differences lie.

Cultural Comparisons

Many tensions are rooted in the different time frames under which journalists and scientists operate. Scientists may happily spend years, or an entire career, delving ever more deeply into a subject—or until the research money runs out. On the other hand, journalists can investigate a subject only as long as their deadlines allow. This may mean an hour, a day, a week, or a month. And as journalist David Helvarg once put it, "Scientists tend to be obsessive-compulsives, while journalists have ADD (attention deficit disorder)."

This constant time crunch is why journalists appreciate a quick overview first as they search for a good story. They are less interested in the details than generalizations and how widely a story applies. Journalists are also always thinking about why their audience should care. Depending on the scope of the outlet they work for, some journalists will be thinking locally, some nationally, and others internationally, so it's important to know which when you are talking to them.

Another cause of conflict is uncertainty. Scientists are comfortable with uncertainty and understand that nothing is ever absolutely "true," but only insofar as it can't yet be falsified, as Karl Popper wrote in his classic, *The Logic of Scientific Discovery* (Popper 1959). This view of the scientific process places everything on a level of uncertainty. Tensions arise when scientists are asked to say how certain they are about their conclusions, and they respond with hedges and caveats. Journalists, policymakers, and the public often misinterpret this to mean that we should wait for certainty before taking action. Absolute certainty, by scientific standards, is almost unattainable. At the very least, it sits at such a high bar that by the time it is reached, it may be too late.

This is the case with climate change. Even though the vast majority of scientists agree on the big picture, discrete areas of debate give the impression that general scientific consensus does not exist. Scientists have to be better at communicating uncertainty or else it will be misconstrued as an excuse to do nothing. As General Gordon R. Sullivan, former army chief of staff, told the *New York Times*, "Speaking as a soldier, we never have 100 percent certainty. If you wait until you have 100 percent certainty, something bad is going to happen on the battlefield" (Homer-Dixon 2007). The same often holds true in environmental science.

Another area of culture clash is credibility. For scientists, credentials matter. Where you studied, who you studied with, and your publication record all mean a lot to your colleagues. Journalists will tell you they are ill-equipped to evaluate the scientific credentials of one scientist versus another. In fact, journalists often look to their "trusted contacts" in the scientific community to help gauge another scientist's credibility. For journalists, credentials are not unimportant, but their real concern is about presenting a range of perspectives.

This focus on perspectives leads to one of the biggest tensions—journalists are interested in you *personally*. They want to know your history, your feelings about a topic, and your motivations. Most scientists want to stick to the facts and the research. You have been trained to be rational and detached—to the point that you write in the passive voice. However, journalists tell stories in the human dimension. People are interested in other people. Scientists are fascinating, even when their research topic might not be. People are interested to know what you do day-to-day, including why and how you do it. Personal details are a "way in" to the story.

Likewise, journalists often like to explore the emotional side of an issue. Near the end of his interviews, while the camera is rolling, Jeff Burnside, a

BOX 3.1
A Few Ideas for Dealing with Uncertainty

David Malakoff

It's the kind of question that terrifies scientists. And it can come anytime from anyone—a journalist, politician, or the lady with the big earrings next to you at dinner. Their eyes narrow like a baseball pitcher about to throw the nasty curve, and then they let loose: "So, how can you be so sure . . . that global warming will melt the ice? . . . that those chemicals are safe? . . . that vaccines don't cause autism?"

How can you answer in a way that acknowledges scientific uncertainty without completely undermining your authority and expertise? Here are a few tips:

Reframe the Question: Focus on what you know, not what you don't know—and the need for specific information.

> **Journalist:** If the forest is cut down, all the bears will disappear, right?
>
> **You:** Well, I can't tell you for sure. But here are some things we do know. We know that bears need food from trees. And my study and others show that, in general, fewer trees lead to fewer bears . . . and where there are no trees, there are no bears. Still, what we need to do is study *this* forest and *these* bears . . . I don't think, based on what we already know, that it would be wise to cut the trees until we have all the facts.

Turn the Tables: Focus on what others don't know, not what you don't know.

> **Journalist:** The company says there's no proof the bears will disappear if it cuts down the forest. How do you respond?
>
> **You:** I think you're asking: "Do we know what will happen to the bears if the company cuts down the forest?" The short answer is that we don't know right now—and neither does the company. I don't think it can guarantee the bears will stay if the forest goes. That's because nobody has enough information to predict how the bears will respond . . . and it seems rash to take an action we can't easily undo until we have that information.

Embrace the Uncertainty: Make it clear you are offering informed opinion—not scientific results.

> **Journalist:** What will happen if the company clears the forest?
>
> **You:** Well, studies are underway but it may be difficult to get enough information in time. So, as a scientist, I can't definitively answer your question yet. But, as a pretty informed citizen, I can give you my opinion. In this case, I think the long-term risks of clearing the forest outweigh the short-term benefits.

BOX 3.1
Continued

Last Thoughts

Don't feel forced to predict the future. Instead, think of yourself as the Weather Channel: tell people that given what you know, and past experience, you think there is a small, medium, or big chance of rain, snow, or wind. That way, you acknowledge the uncertainty—and still give an answer that people can use to help shape their own opinions and actions. My guess is, you'll still have a better batting average than the weather forecaster.

—David Malakoff is an independent journalist and science editor. He was a former science editor and reporter for NPR and *Science* magazine.

---

television reporter for NBC TV (WVTJ Miami), often leans in to ask his favorite question, "And how did you *feel* about that?" This sometimes stops the scientist dead, but other times the question can trigger a heartfelt response. Once, when interviewing a scientist who researches potential cancer-curing compounds in the rainforests of Costa Rica, Burnside asked if anyone close to him had suffered from cancer. The scientist paused and then answered, "My father died of it." Burnside followed with, "And do you ever think about your father when you are working in the rainforest?" "Yes . . . I think of him every day," replied the scientist. This powerful personal exchange made the story for Jeff and his viewers.

In summary, the table below illustrates the differences between these two cultures that often lead to conflicts.

Differences between Science and Journalism

Science	*Journalism*
Slow and ongoing	Deadline-driven
Evidence first	Conclusions first
In-depth	Quick overview
Uncertainty	Certainty
Specifics	Generalizations
Credentials	Perspectives
Rational	Emotional

Whose Voice Is in Your Ear?

One of the biggest differences between the cultures of scientists and journalists is who looks over their shoulders—in other words, whose judgment they
value. In interviews, scientists often answer questions as if they were talking to
other scientists. This is because they can hear the judgment of their colleagues
ringing in their ears: "Did I forget someone I should have acknowledged?
Did I sound silly? Will I have offended anyone?" An important part of becoming an effective science communicator is to stop talking to your peers.
Start thinking about what the journalist and his or her audience needs from
you: clear, concise, conversational answers.

Journalists have their own little devil on their shoulder whispering in
their ears. They are thinking about their audience but, perhaps more important, they are thinking about how their editors will react. If a story is vying
for front-page placement in the paper, or online or billing at the top of the
TV news hour, an entire chain of editors may weigh in.

The journalist has to sell his or her story. The competition is fierce. Kenneth (Ken) Weiss of the *Los Angeles Times* covers environment and aims for
front-page stories whenever he can. "It's a challenge," he admits. As he explains in our workshops,

> Editors are a pretty good proxy for the public but they have a shorter atten-
> tion span and can be a whole lot crankier. The stories I write about the en-
> vironment have to compete with everything from stories about the war and
> terrorism to the latest antics of celebrities stopped for drunk driving.

Hence, the more scientists can help journalists by cutting to the chase and answering the nagging question "why should we care?" the more likely the
story will make it past the editors.

You've Got It All Wrong: Some Big Misperceptions

The culture clash between scientists and journalists leads to misperceptions.
In conversations with both sides, the following themes come up time and
again.

Scientists' Perceptions of Journalists

"The Media" Are All the Same

Scientists tend to refer to "the Media" as a monolithic entity, as if journalists were all the same. Nothing could be further from the truth. There is just as an elaborate taxonomy of journalists as there is of scientists. A print journalist for the *San Francisco Chronicle* is utterly different from a television journalist for CNN or an online reporter who blogs. This goes beyond the fact that the medium dictates how the story is told. Just as important, journalists bring their own interests, geography, background, and context (as well as those of their organization) to every interaction with a scientist. Reporting is different from one country to the next, as well. Hence, scientists would do better to consider individual journalists and their venues. See the essay by British journalist Tim Radford as an example. There is no such thing as "the Media."

BOX 3.2

A Journalist's Perspective on the Two Cultures

Tim Radford

Scientists and journalists both are interested in providing accurate and reliable information to a larger society. But what matters most to a scientist may not be quite the point that a journalist will seize upon. Don't be surprised: you know your own field only too well. You know it so well that you might not see the big picture, the salient fact, the thing that will seize and hold the interest of the uninitiated.

Yes, there is a danger that if you talk generously, unguardedly, and frankly, you might misspeak, or be misheard or even misquoted. But the more forthright and simple your choice of words, the less likely it is that this will happen. The more openly and helpfully you talk, the more likely it is that the attentive reporter will pick up the error and—if it is obviously an error—silently correct it, and the more likely it is that in the course of speaking, you will deliver one of those graceful, telling sentences that get quoted everywhere.

Scientists who try to avoid public debate tend to get a very bad press. When they actively involve the media in the discussion of science, they tend to get very good, sympathetic coverage. As a consequence, certain scientists are sometimes condemned by their peers as "media tarts." They are the scientists who tend to appear in media repeatedly.

BOX 3.2
Continued

You may already have worked out why. We prefer persuasive people who answer the phone promptly, and then answer the question convincingly. And we are really bad at consulting those people who say "No comment," or, even worse, "Well, I'm hardly the right person to ask old boy, you should get in touch with Professor X, he's away right now but I'm sure he'll be delighted to help when he gets back," or, worst of all, "Well, if you'd read my article in *Nature Geophysics* seven weeks ago, you'd know that your question is simply ignorant. . . ."

Perhaps journalists should spread the net of questions far wider. Maybe, but you try spreading your net wider at 6:30 p.m. with 800 words to deliver by 7:15 p.m. But that's not the real reason why journalists prefer coherent and articulate scientists to evasive and dismissive ones. My point is that "no comment" should never be an answer. What is the point of research if it doesn't end up with answers? What is the point of your own research if you would rather defer to someone else? And what is the point of telling your interviewer that he is an ignorant little twerp? Reporters know that they are ignorant. That is why they ring up someone who they hope does know the answer and is prepared to tell it.

I don't expect my scientific source to be the last word on a subject. I am a science reporter, and my questions are almost always about science that has just been published, and a finding that is therefore likely to be startling, contentious, alarming, or even amusing: in a word, sensational. Why else would I ask questions? Why else would anyone want to hear your answers?

—Tim Radford was science editor of the *Guardian* until his retirement. He continues to report for the *Guardian*.

They'll Misquote You

Scientists often accuse journalists of misquoting them, as if it were intentional or malicious. They don't realize the seriousness of this allegation. Most journalists are scrupulous when it comes to direct quotes. Sometimes a problematic quote is correct but feels embarrassing once it is in print because it is what you *said* but it isn't what you *meant*. So while misquotes can—and will—happen occasionally, there are things you can do to help make sure what you mean to say is what comes across. Knowing exactly what message you wish to convey puts you in a better position to communicate it clearly. If you have stumbled in an interview, check in with the journalist to see if you were clear, but be sure to do it in a way that comes across as helpful rather than controlling. Much like scientists, journalists want to get it right—and you can help them.

They Don't Care about Accuracy

A journalist's most important currency is his or her reputation, which hinges on accuracy. Journalists' credibility with their editors, colleagues, and sources depends on it. If a journalist makes an error, most media outlets will publish a correction, often somewhere prominent. Online, the correction often goes at the very top of the piece and stays there in perpetuity for the world to see like a badge of shame. When it comes time for journalists' evaluations, the editor will tally the number of corrections and consider them an important part of their overall performance. Scientists can help journalists with the challenge of getting it right by being accessible and available to talk again after the interview for further questions or fact-checking.

They Don't Know Anything

Many of the best journalists don't have formal training in their topic, but when they are allowed to focus on a "beat" they can become true experts. For example, Andrew (Andy) Revkin who writes the Dot Earth blog for the *New York Times* is deeply knowledgeable about climate change. Ken Weiss of the *Los Angeles Times* knows a great deal about oceans. They have gained profound knowledge much as you have—by reading and discussing their ideas with researchers. As a result, they can investigate and debate these subjects to a degree that scientists find challenging.

Some journalists do have formal training in the sciences. For example, Natasha Loder of the *Economist* has a PhD in entomology. David Kestenbaum

Non Sequiter © 1993 Wiley Miller. Dist. by UNIVERSAL UCLICK. Reprinted with permission. All rights reserved.

of National Public Radio has a PhD in physics from Harvard. But Natasha covers technology, environment, and economics, while David reports on everything from Medicare to the recession.

Journalists are smart, savvy people who can quickly get to the heart of a problem and make sense of it for their audience. In fact, being too educated on a topic can be a liability when one's job is to translate the subject to a general audience. Many journalists will not tell a scientist if they do have expertise because they want *you* to do the explaining.

Journalists must have the aptitude and ability to range far and wide in their reporting. Cory Dean, former science editor of the *New York Times*, puts it in perspective when people suggest that all science journalists should have PhDs. Her response: "In what?"

They Sensationalize!

It is sometimes true that journalists sensationalize. Journalists are looking for a good story that will capture the interest of their audiences, and sensation sells. A loud headline may not necessarily always be a bad thing. It can attract people to something they might not otherwise pay attention to, and can be a way in to learning more about a subject. However, the level of sensationalism depends largely on the outlet's style and, to some degree, the style of the journalist. Think of the differences between Fox News and the *NewsHour with Jim Lehrer* on PBS, for example. Science journalists generally avoid overhyping a story to protect their credibility as well as that of their media outlet. And just as scientists should avoid journalists or media outlets prone to excessive exaggeration, most science journalists avoid sources that are known to overstate their findings.

Okay, so what do journalists say about you? The following are common themes.

Journalists' Perceptions of Scientists

You Love Caveats

Scientists use caveats in their efforts to be absolutely accurate and to make sure you do not overstate your research findings. For the journalist or general public, however, the main points can get lost in the qualifiers. Caveats undermine clarity, which is what journalists seek. Make the statement as clearly and simply as you can, then add the caveats if you must. Turn your typical response on its head. If a journalist suggests a headline or summary statement, instead of asking "Is this statement true in every circumstance?" begin instead by asking, "As presented, is this false?"

You Are Overly Interested in Process

Remember to mimic the basic tenets of journalism: begin with the punch line and then draw the journalist into the fascinating details once he or she is "hooked." The journalist is dying to know what you actually found at the end of the day. That discovery or advancement will often determine if the research is interesting enough to warrant a general-interest story. Yet many journalists *are* interested in the process. NPR's Christopher (Chris) Joyce loves stories about how scientists figure things out, especially when told like an unfolding mystery.

You Lack a Bottom Line

This isn't always the case, but often the scientist doesn't state his or her conclusion clearly enough for the journalist to understand. Or perhaps the scientist takes so long to get there that the journalist loses interest along the way. Journalists typically read scientific articles by starting with the title and abstract, then skipping ahead to the conclusion, which usually contains what they want to know. A journalist wandering through a poster session at a scientific meeting will often walk with eyes trained to the bottom of the poster. If the conclusion gets his or her attention, then he or she will stop and read for more details. For journalists it's usually the concluding paragraph of a scientific paper that's most interesting. It is important to think about pulling these conclusions up front when you are talking to anyone other than a colleague.

You Speak Jargon . . . and Alphabet Soup (Acronyms)

Nobody likes to look stupid. While journalists usually don't mind asking for a definition here or there, take the time to trim as much jargon as you can. For example, don't just say benthic, either say "bottom-dwelling, which we call 'benthic,'" or better yet, just say "bottom-dwelling." It is sometimes okay to use the term, but define it. Otherwise you may be several sentences ahead while the journalist is still puzzling over what that word meant. Also, you don't want the journalist to remember the definition instead of your main point. The issue of language choice is always a topic of debate. In fact, there may be a slow evolution as words that were formerly considered jargon go mainstream. Think of terms like biodiversity and greenhouse gas that only recently found their way into news stories. Nevertheless, err on the side of being plain-spoken. Journalists appreciate it.

The relationship between scientists and journalists can be combative or synergistic. Journalists typically want to help scientists do a better job of communicating their science. They strongly believe that society can benefit from better information. But in order to achieve the best results, each party must better understand the other's culture and constraints. Scientists often say to me, "Why do *we* have to change? Why can't the journalists accommodate us—or can't they at least meet us half way?" The short answer is, they don't have time. Reporters toil under the tyranny of crushing deadlines. That is their limiting factor. If you can help the journalists get what they need before deadline, they will be extremely grateful and the story will be better for it. They may well come back to you next time.

The Bottom Line

The tensions between the worlds of scientists and journalists will likely always exist. But by understanding each other's culture, and trying to accommodate each other's needs, you can find common ground. By being more accessible—in language, attitude, and availability—scientists can go a long way toward more frequent, higher-quality science coverage in the news.

Chapter 4

TELL ME A STORY: WHAT JOURNALISTS WANT FROM YOU

The secret of being a bore is to tell everything.

—Voltaire

Scientists talk about science, but journalists and the rest of the world want to know, "What is the *story*?" Our attraction to stories seems hard-wired into our psyches—they help us make sense of the world. People have told stories since the earliest days around the campfire. Well-told tales appeal to our reason and emotions in a way that theories and generalizations typically do not (Egan 2004).

Journalists are professional storytellers. They find story ideas everywhere: in casual conversation, over late-night drinks at conferences, chats in passing, e-mail tips, blogs, Twitter, and checking in with their scientific contacts from time to time.

They talk about having a nose for news. Many journalists tell me they experience a little electrical jolt—"aha"—when they smell a story worth telling. I get a prickling on the back of my neck. It might be the sort of feeling scientists get when you look at your data and see something revealed.

But you need to find the *story* in what you want to convey. Stories are quite different from topics or issues. For example, climate change is an issue, but composting as a strategy to remove carbon dioxide from the atmosphere, which is research that Whendee Silver of Berkeley is doing, has the makings of a good story. Who knew that composting is a way that individuals can do their bit to nibble at the climate problem rather than adding to it by dumping their waste in landfills? It's not only surprising, but it offers people a chance to actually do something. Chris Joyce pricked up his ears and made plans to visit Whendee at Berkeley when he heard about this at a communications workshop. "That's a great story," he said. 🔨

Stories are the currency of journalism. New, surprising, and relevant—these are some of the most common elements of any good story. But every journalist has his or her own particular tastes. I asked some well-known journalists to summarize what they think makes a good story in a just a phrase or two:

Jeff Burnside, NBC Miami News: A "wow factor," affecting people's lives, characters, visuals of course.

Cory Dean, *New York Times*: Human drama, I love stories that say some known scientific fact is wrong, also quest stories.

Juliet Eilperin, *Washington Post*: A larger trend, something that has impact—something that matters, or gives insight into scientific or political process, gee whiz can be very interesting . . . oh and images and graphics.

Tom Hayden, freelance writer: Explanation of a common phenomenon that was previously unexplained, something concrete—something I can see or touch or smell—concept only is difficult to sell.

Chris Joyce, National Public Radio: Novelty, the unexpected, passion, irony . . . a story to make my editor love me.

Natasha Loder, *The Economist*: Specifics to back up big ideas, counter-intuitive.

Lance Orozco, KCLU Local NPR News: Interesting to Joe Six-Pack, with a take-away message, transcends the audience to have universal appeal, audio and pictures.

Ken Weiss, *Los Angeles Times*: New, "shows rather than tells," reveals fundamental truth, interesting characters, interesting places.

If you pay attention to an individual journalist's work, you'll find his or her preferences expressed in his or her stories. Analyze them. This is an instructive exercise on its own, but it is especially important if you are interested in ever having that person cover a story related to you and your work (see chapter 11).

What Makes a Good Science Story?

> My role as a journalist is to find what's new and different in
> the world, what should you think about it, and maybe what
> should you do about it.
>
> —Andrew Revkin, *New York Times*

If you think of your work as a story, you will find it easier to present it to
journalists that way. As a result, you'll find them much more interested than if
you had simply unloaded a lot of data. A good story will have one or more of
the following qualities:

Novelty

Tell me something I don't know. Tell me something surprising. Nearly all journalists—especially science reporters—are looking for what's new and surprising. But reporters and other nonscientists often need help figuring out what is
truly new, since they are not attuned to the history and progression of your
particular field and are not interested in incremental advances. Be sure to explain how the work advances on previous knowledge and why it matters.

While the bigger story may not be new to your colleagues, it will be for
many people, and you do a science a service by telling the broader story. Stories can include previously published work that went unnoticed outside of
scientific circles. When a journalist can tell a good story that puts it all together, suddenly it's news. You can practice explaining what is new about
your work and, most important, why it matters. This will really help you
when the time comes to describe it to a journalist, or even at your next cocktail party.

Passion

Not everyone is interested in science, but people *are* interested in other people. Journalists love stories about scientists and their passion for their work.
One of Chris Joyce's favorite examples is a story he did about a young bird
researcher who was helping some colleagues by holding down an alligator in
the back of a truck (Joyce 2002). They were taking it back to the lab for

blood samples. Joyce loved her enthusiasm recounting her tale because she described the experience like she was talking to a friend, "Someone was driving way too fast, I can tell you that. . . ." While she was careening around corners, trying to keep her position astride the gator, she stared closely at its face and noticed it was covered in small dots. She went home and looked them up "just out of plain scientific curiosity," but found nothing describing them. This led her to doing research that revealed the dots to be sensory pits. Eventually her curiosity resulted in a *Nature* paper. It's a great story, and you can listen online. 🖱

People are curious to know what it's like to be you. Why are you willing to spend years studying one particular thing? What did all of that dedication reveal? Character-driven narratives allow journalists to delve more deeply

---

BOX 4.1
Why Be a Character?

Michelle Nijhuis

Scientists love to erase themselves. Your study site "was chosen," the data "was collected," and the conclusions "were drawn." We journalists, on the other hand, have the annoying habit of describing you—the human being—as you choose, collect, conclude, and otherwise practice science. We like to turn you into what we call a "character."

When we portray scientists—as I have—being shipwrecked with only a chocolate cake as sustenance, engaging in a battle of wills with a crow, toasting colleagues with tequila shots after a day in the field, or reminiscing about childhood bug-eating contests, it may seem that we're simply out to embarrass you. But for science journalists, such details—and the characters they bring to life—are indispensable tools.

While you may spend most of your days surrounded by other scientists, many of our readers don't know any professional researchers personally. They may believe the popular stereotypes of scientists as forbiddingly cerebral and hopelessly unfun. Even though they're often fascinated by scientific subjects—wildlife, astronomy, geology, genetics, you name it—their assumptions may leave them intimidated by, or averse to, scientists and science.

One of our jobs as science journalists is to make science more accessible to such readers, and one of the ways we do that is to describe scientists as they really are: perceptive, passionate, fallible individuals with stores of important knowledge and a huge variety of motivations, quirks, and talents. In short, irresistible characters.

So when I'm sitting in the passenger seat of your pickup truck, or hiking into the backcountry with you, I'm listening to what you tell me about your results and their

---

BOX 4.1
Continued

implications. But I'm also getting to know you. I'm listening to the stories you tell about your childhood, your early days as a scientist, and the path that led you to the study at hand. I'm listening to the jokes you make, the conversations you have with your colleagues, and your reactions to daily victories and defeats. I'm noticing what you're wearing, what you're eating for lunch, and what you like to listen to on the radio.

I don't put all these observations into the final story, but I write them down in my notebook anyway because I know that some will help me introduce you to my audience as a complex, interesting human being. The more intrigued my readers are by you as an individual, the more likely they are to set aside their assumptions about scientists—and the more likely they are to read, understand, and remember the story of your work. Journalists use characters for a broader reason. Stories are the most effective containers and carriers of information we have, and to tell stories about science, we need to tell the stories of the people who choose to make it their life's work. After all, a story without characters is like a stage set without actors—tantalizing, maybe, but far from the whole show. And while science journalists have managed to tell a few amazing stories starring molecules, potatoes, and other nonhumans, good storytelling usually requires people in the lead roles.

It may seem like a bother to allow a reporter to visit your lab, or accompany you to your field site, or barrage you with seemingly irrelevant questions. You may wonder why you can't just send us your latest paper and be done with us. In fact, you can—you're certainly under no obligation to share personal information with a reporter. But to do our jobs well, we almost always need more than your results. We need you. So the more you're willing to tell us about yourself, the more you can help us understand why and how you do what you do, the more engaging the resulting story will be—and the more impact your work will have.

—Michelle Nijhuis writes for a wide range of publications including *High Country News, Audubon, Smithsonian, Orion, National Geographic,* and the *New York Times.*

---

into the juice and joy of science instead of just recounting the facts. The audience is drawn to the person doing the work—you.

Mystery

A journalist's goal is to have the audience hanging on each word, dying to know what happened next. Science is all about asking questions and solving puzzles, and it is very appealing to journalists if you can talk about your work as a mystery. What question are you trying to answer? What are the plot twists? Who are the characters? Journalists are not only interested in published papers and results, they also enjoy the journey of discovery along the way, especially if the story has a narrative thread. Unsolved mysteries can be compelling precisely because they haven't been completely solved yet. It's nice, of course, for stories to come wrapped in a neat and tidy bow in the form of a well-founded answer, but when the question is intriguing, the pursuit of an answer can be enough.

Pat Conrad, a professor of parisitology at the University of California Davis, is investigating the mystery of the disappearing sea otters, a story that continues to unfold. Despite federal protection since 1977, southern sea otter populations have not recovered. Conrad's team discovered that sea otters were dying as a result of a brain disease caused by a single-celled parasite, *Toxoplasma gondii*. This microscopic killer is shed in cat feces and kitty litter, and then carried into the ocean by freshwater runoff. "This story had all the critical elements to attract media attention," Conrad says, "mysterious disease, charismatic creatures (sea otters and cats), violent death, and sex (of the parasite inside the cat)." (See Box 8.1 on page 105)

Adventure

Your time spent in the field might seem workaday to you, but compared to the average person, you live a life of high adventure: exotic locales, wild animals, extreme weather, and triumphs over difficult circumstances. Scientists are so focused on the data that you sometimes don't realize how exciting all that backstory can be. Journalists enjoy witnessing the process of collecting information and how scientists get up close and personal with their subjects, sometimes under extreme conditions. They have an expression, "when something goes wrong it goes right." For example, being chased into a tree that is too small to support you by a rhino you are trying to study (true story) is a great introduction to a story about rhino research. Once you get your audience's attention you can get them interested in the details.

Overturning Conventional Wisdom

A sure way to get someone's interest is to present them with the unexpected. As *New York Sun* editor John B. Bogart said, "when a man bites a dog, that is news" (Mott 1950). And when new information refutes common knowledge, that's a story. For example, *Archaeopteryx*, believed for 150 years to have been the first bird, was probably only a feathered dinosaur (Erickson et al. 2009).

Results that overturn common knowledge or common sense are particularly appealing, as is evident in most health and science coverage. Unfortunately, scientific whiplash can lead to public confusion and frustration. When "experts" can't seem to agree—is coffee good or bad for me? should I eat farmed salmon?—no one knows what to believe. Then the public starts to ignore the debates, assuming it too will be debunked tomorrow.

When you are dealing with this sort of issue, don't ignore it or pretend it doesn't exist. Be prepared to acknowledge recent or repeated reversals in your field, whether they are real or perceived. Practice explaining. You might want also to focus on why you personally were surprised by your findings, rather than attacking other people's assumptions. And it's good to remind people that this is how science progresses.

Conflict and Controversy

While it may be distressing to scientists who would rather avoid public disputes—it's true—conflict and controversy sell. It isn't that journalists are always digging for dirt to sully your reputation or make you look like a fool. It's that the best stories usually involve some sort of conflict, and sometimes a resolution. Without it, a story falls flat. If Odysseus simply sailed around the Mediterranean without travails or mishap, do you think Homer's stories would have had the same shelf life? We are a culture accustomed to conflict and even addicted to it. We want to find out: what happened next? Who won?

Unfortunately, the media's appetite for discord is fertile ground for professional doubt-makers. This is why it is important for you to be able to explain how there can be overall scientific consensus while discrete areas of argument still exist. A troubling example is the broad scientific consensus that human-caused climate change is happening, even though debates rage on to the point that public confidence is waning. So be sure to say exactly where there is consensus—do not assume people know this.

Just Plain Cool

Journalists and their audiences like stories with intrinsic appeal—some call it the "wow factor." One example is how bees' knees grip 🖱, which was covered on *Quirks and Quarks*, the Canadian Broadcasting Company's science show (McDonald 2009). This nationally broadcast program aims to surprise and intrigue listeners. As Jim Handman, the show's senior producer, explains, "Since scientists do most of the talking for this program, the test that we apply is not only how interesting the topic is, but how interesting is the scientist?" Before a scientist is interviewed he or she is screened in a preliminary phone call. If you can't talk about your research, it's no go. Enthusiasm is critical. If you aren't excited about your own work, why should anyone else care?

Think about your research. Are there any superlatives connected to your subjects or findings—biggest, oldest, fastest, coldest? Do you investigate parasites, Amazonian rainforests, or deep-sea creatures? Do you study sex, death, disease, or conflict (competition)? Do your methods include lasers, remote control vehicles, or toxins? These are just some obvious starting places, but look at your work with fresh eyes. What would someone who knows nothing about it find weird, fascinating, or just plain cool?

BOX 4.2
Tell Me a Story

Jim Handman

There's a well-known tale, told by a frustrated public affairs officer for a leading American university, about trying to get a senior scientist to describe his latest groundbreaking paper so that she could write a news release about it. But he seemed incapable of translating his complex experiment into layman's terms. After several unsuccessful attempts she finally told him to describe it to her as he would for his mother. "Madam," he said severely, "I do not discuss my work with my mother."

This is the challenge that we science journalists face daily. And nowhere is that more critical than in my field of radio broadcasting. We do not have the luxury of translating a scientist's words into layman's terms, as my print colleagues do. They need only find a twenty-five-word quote to insert into their summary of what the scientist told them. After all, in a twenty-five-minute interview, you can always manage to find twenty-five words that make sense. But in radio, we have to actually find scientists who can summarize their own work in a clear, concise, and conversational manner—what we call "the three Cs" of broadcast writing. And that turns out to be a far greater challenge than you might imagine. Scientists, as a breed, seem genetically incapable of using simple language when complex jargon will do. Why talk about an animal's form and structure when you can just say morphology?

BOX 4.2
Continued

So we tend to tell the scientists we interview every week the same thing that the poor frustrated public affairs officer said. We tell them to pretend they're at a cocktail party, where a stranger asks them what they do for a living, and what their latest research is about. They may lack any background in or knowledge of your specific field. But they do know a good story when they hear one.

Scientific research can provide vivid, engaging, and captivating stories. In fact, science papers actually follow a fairly predictable "narrative arc"—as we call it in journalism. There is a beginning (what you set out to find), a middle (how you did your experiment), a turning point (what you discovered), and an ending (your conclusions, or what it all means). You just have to make it clear, concise, and conversational. How hard can that be?

And that doesn't mean you have to "dumb it down." Your cocktail party strangers may not be scientists, but if they can grasp the finer points of economic policy or Middle East politics, then they can probably get the gist of habitat loss or speciation—as long as you make it a story that they can follow. And don't leave out the really cool bits. Your yarn about camping on the Arctic tundra in the middle of a blizzard may not be part of your published paper on the predator/prey relationship of the lynx and the snowshoe hare—but it's a damn good story, and actually gives people a sense of what scientific field work is all about. Don't be afraid to tell us how surprised you were when you discovered that the albatross couples tending the nests were both females—it's refreshing to hear a scientist express an emotion.

As I say to every scientist we interview, this is the easiest test you'll ever take—since you wrote the questions yourself. We're just going to ask you about your work—the work you've spent years researching. All you have to do is tell me about it. Just tell me a story.

—Jim Handman is the senior producer of Canada's national radio science program, *"Quirks and Quarks,"* broadcast on CBC Radio.

It's Not Necessarily the Whole Story

This is a tough one. Journalists are usually looking for a "bite" of the story rather than the entire history and chronology of any particular scientific discovery. Thinking in terms of bite-sized pieces can go a long way toward helping journalists find a story that will work for them, their editors, and their audiences.

Which Stories Make the News?

News editors typically focus on daily events, not long-range issues. Complex subjects get ignored if they are not specifically linked to daily life. Environmental issues suffer because they are often about gradual trends like species decline or habitat degradation, instead of breaking events. They tend to present complex problems that are difficult to explain briefly and may not have a local impact. The link to ourselves can be hard to make because the impacts may not be seen or felt immediately.

So, ecological issues often don't pass the litmus test for news. Why not?

- Many stories are a gradual trend—for example, species decline, over-fishing, climate warming.
- Issues are abstract, and the losses are not always obvious.
- Issues may be global, and media is often local.
- The best examples are difficult and/or expensive to capture visually.
- There are often no clear-cut villains, or the enemy is us.
- Problems are complex and hard to explain briefly.

Media coverage can help float environmental issues to the top of public and political awareness but it may give a skewed impression. For many years, repeated public polls showed that the people believed that the biggest issue facing the oceans was oil spills (Belden et al. 1999). In fact, overfishing, climate change, and ocean acidification are larger, more serious threats. Yet oil spills repeatedly trumped all in the public's mind, because their impacts are visible and well reported. Consider the 1989 Exxon Valdez oil spill in Alaska's Prince William Sound and the 2007 spill in San Francisco Bay, for example. These events triggered waves of coverage because they had all the right ingredients for breaking news. There were compelling pictures of devastation. As journalists say, "if it bleeds it leads." There were villains to blame: a drunk sea captain in the case of Exxon Valdez and an inattentive one who struck the bridge in San Francisco. Cute cuddly animals were affected, including sea otters and other marine mammals, birds, and fish. Heroic clean-up crews struggled to save them. The problem was easy to understand: the natural environment was devastated and so were people's livelihoods; animals died. Above all, the visuals brought it home and made it real.

Other environmental issues are not so immediate or obvious. For many years, Andy Revkin, a pioneer in covering climate change, fought an uphill battle to get his climate stories into the news. The accumulating evidence convinced scientists, but the urgency was harder to convey to the public. As Revkin explains, "You will never see a headline that says 'Climate change broke out today.'" His main challenge was that the developments were incremental, and predictions called for the ill effects not to register until sometime in the future.

"Scientists have the future in their bones, while the traditional culture responds by wishing the future did not exist," writes Peter Dizikes in an essay about C. P Snow's book, *The Two Cultures*, about the schism between science and society (Dizikes 2009). This is an enduring problem.

What Doesn't Make News?

Understanding what doesn't make news is as important as having a sense of what does. In general, anything that can be perceived as self-serving or self-promoting does not make the grade. This includes many things that institutions might like to see as news:

- Faculty and staff appointments
- Routine awards and grants
- New facilities
- Incremental findings

The exceptions to this are small community venues that cover "insider baseball," such as campus newspapers, and media that focus on local news. Appreciating the appropriate scale and fit of your story can best be learned by paying attention to each outlet you might like to target and noticing what gets covered by whom.

Breaking News or Feature Story?

The first fork in the journalists' taxonomic tree is whether a story is breaking news or a feature. Breaking news stories in science are relatively rare. The bar is high. It takes a major new discovery or often one of the negative triggers: crisis, conflict, criticism, corruption, or catastrophe. Science news stories tend to be relatively short and stick to reporting the facts.

However, the potential for science and environmental feature stories is limitless and can take many shapes. Features *can* sometimes address incremental environmental issues.

Ken Weiss of the *Los Angeles Times* wrote a Pulitzer Prize–winning series in 2007 called Altered Oceans, which detailed the environmental creep of pollution and the "rise of slime" (Weiss and McFarling 2006). "Too many environmental stories are about future or conditional threats," said Weiss.

> To give Altered Oceans some currency we looked at how the rise of slime was affecting people today. I decided the way to tell the story was not to tell it at all, but to show it. I focused on the blowback coming on shore, and used that to tease out the stresses on the oceans.

Weiss's five-part series was based on interviews with hundreds of scientists but the final stories were about people: fishermen suffering rashes or catching nets full of jellyfish instead of shrimp, and families who moved to remote barrier islands in search of paradise but instead found themselves breathing the aerosolized toxins that floated on the sea breeze.

Feature stories can take many shapes. They can range from short blurbs to in-depth explorations. Full-length features require more sophisticated narratives and storylines. This is where your creativity in helping identify the story can pay off.

Know Your Media: Four Perspectives

All journalists share the same "wish list" when it comes to scientists: they want you to be clear, concise, and colorful. However, they have some specific and very individual needs depending on whether they do their reporting online, in print, on the radio, or on television.

While it's tempting to think of "the media" as one audience, we're really talking about many different audiences, each with its own needs. Just as you would approach, say, a policymaker from California differently than you

BOX 4.3
What Makes a Feature Magazine Story?
A Perspective from Smithsonian *Magazine*

Laura Helmuth

People read magazines for a lot of different reasons: to improve their minds, to accumulate amusing cocktail-party banter, to distract themselves while waiting for an impending dental exam. But most of all they want to be entertained.

Smithsonian magazine's mission is consistent with the mission of the Smithsonian Institution, which comes straight from James Smithson's will: "the increase and diffusion of knowledge." At *Smithsonian* magazine, we diffuse knowledge in part by making our stories a pleasure to read.

Very few of our 7 million or so readers have any background in science, and many did not go to college. They may subscribe because they like our history or travel or art stories; they don't necessarily come to us for science. The topics that appeal to them most tend to be archaeology and research on charismatic megafauna (readers love elephants). But they'll read and enjoy stories about almost any scientific topic if the stories are told right.

We know what sorts of ingredients go into a good story: interesting characters, photogenic settings, dramatic action, coherent narrative. The trickiest thing is probably finding a story that has a beginning, a middle, and an end.

Science, of course, is a neverending story. The best scientists pick important questions, answer them, and immediately use those answers to pose a new set of questions. But the best magazine writers and editors are always looking for a satisfying ending. That's one of the elements that separate magazine stories from news stories. Magazine readers expect a more complete package, with more answers than questions.

Being part of a *Smithsonian* story is a big commitment. If you're a major character in the story, a writer will spend at least several days trailing after you as you do your research. They'll take notes, push you to explain your work using metaphors, get in the way, and ask dumb questions (which is sometimes but not always because the writer wants you to say something more colorfully or simply). And that's just the start. The writer, who is usually a freelancer, will send the story to an editor (that's me), and I'll have many more questions and requests and revisions, which will require the writer to have more phone conversations with you. After I and at least one other editor are satisfied with how the story reads, it goes to a fact-checker, who will call you and ask the exact same questions that you already answered when the writer asked them several months earlier.

It is unsettling to let someone else tell your story—you are the expert on your own research and life, and any story-sized piece of it necessarily omits important points. You may feel that your work is being dumbed down, and it is.

BOX 4.3
Continued

What do you get out of it? Well, you're diffusing your knowledge. You're reaching people who otherwise have very little science in their mental diets. And the end product—a magazine with gorgeous photos and an elegant layout and what we hope is compelling text—is satisfying as an object, as something that people want to hold and have in their homes, something that inspires them. And if we've done our jobs right, readers will want to tell other people—in their own words, or by lending their friends a copy of the magazine or forwarding them a link to our website—your story.

—Laura Helmuth is a senior editor for *Smithsonian* magazine.

---

would approach one from Kansas, it's important to customize your communications to suit individual media and their audiences.

In Boxes 4.3–4.6, I have invited a few top science journalists from various niches in the media ecosystem to tell you what *they* deem important in a story. What do they look for in their stories and what do they want from *you*? Notice the similarities and the differences.

Laura Helmuth is a senior editor for *Smithsonian* magazine. Dawn Stover is a longtime editor of *Popular Science* who also freelances for a wide range of media, from the *New York Times* to MSN.com to *Outside*. So, she knows well the importance of considering your audience.

The Economist

Decision makers the world over read the *Economist*. The stories do not have bylines because the magazine's strong editorial direction means it "speaks with one voice." A paper version is published weekly and is supported by a sophisticated online presence, including an environmental blog called Green.view. The website provides much more space for stories than the highly competitive and limited print version. If your science is reported in the *Economist*, the impact is huge.

When *Science* published a paper titled "Can Catch Shares Rebuild Global Fisheries?" (Costello, Gaines, and Lynham 2008), a story, editorial, and podcast by the *Economist* (Anon. 2008b; Anon. 2008c; Anon. 2008d) prompted people around the world, including decision makers, to contact authors Chris

BOX 4.4

Know Your Audience—Advice from an Editor and Freelancer

Dawn Stover

I was sitting at an airport gate recently when I noticed a man reading the latest issue of *Popular Science* magazine. I couldn't help staring at this mythical creature, one of the 6 million readers I had written so many articles for, but had rarely seen in the flesh. I thought I knew a lot about him: He was forty-three years old, had spent at least a year in college, owned his own home, and had a household income of almost $73,000 a year. He was tech-savvy and curious about how things work. He liked hard, shiny objects.

Of course, that perfectly average reader exists only on paper. Even so, every magazine has a unique identity that is inextricably linked to the characteristics and tastes of the people who read it. Knowing your audience is essential for editors who create magazines, and for writers who pitch article ideas to those editors. But too few scientists understand that knowing your audience is equally important for anyone being interviewed by a writer, radio or TV reporter, filmmaker, blogger, or journalist of any kind.

The first step in knowing your audience is to learn as much as you can about who is reading, watching, or listening to the media outlet where your words will be published or broadcast. Most outlets collect information about age, gender, education, affluence, and other demographics of their audience. Sometimes you can find this information online in a media outlet's "about us" section. Another good place to look is in the "advertising" section, where many publications post media kits that describe their readers to potential advertisers.

Most media outlets also have a mission statement or motto, and usually it is mentioned prominently as part of their branding. For example, on the cover of every *Popular Science* you'll find the words "The Future Now" printed inside a circle next to the name of the magazine. The editors want everyone to know that this is a magazine for people who are interested in cutting-edge technology.

The best way to learn about a media outlet's audience is to join that audience, if only for a brief time. If it's a newspaper or magazine, pick up a recent copy and skim it. If it's radio or TV, watch or listen to the program. Pay special attention to work done by the person who will be interviewing you. You'll quickly get a sense not only of the educational level of the audience but also of the outlet's "voice"—its style of writing, tone, and attitude. One publication or program may be serious and stodgy while another is youthful and irreverent. One may publish long, technically detailed articles while another sticks to short, simple pieces. Recognizing these differences will enable you to adapt your own voice accordingly.

Also be aware that differences in format reflect different types of audiences. If you're going to be interviewed by a radio reporter, for example, you can expect his audience to be interested in the sounds associated with your work. If the reporter is writing something for a website, his readers don't have patience for long-winded explanations.

BOX 4.4

Continued

A reporter is more likely to include your quotes, ideas, and anecdotes in his article or program if he finds them interesting himself. Remember, though, that the reporter is a talented middleman who isn't necessarily representative of his readers. He might be younger or better educated or more urban than his readers, for example. He might even forget to ask the most important questions of all: "Why should my readers care about this? Why now?" In the end, the reporter is not your audience. It's that guy in the airport.

—Dawn Stover is editor at large for *Popular Science* magazine as well a freelance science writer and editor.

Costello, Steve Gaines, and John Lynham. "I can't tell you how many people saw the study in the *Economist* that didn't see it in *Science*," says Gaines. "It was critical in terms of catching the attention of the policy community."

Natasha Loder is the reporter who covered it. Loder's colleagues playfully call her "The Fish Correspondent" because of her penchant for fisheries stories. A feisty, incisive journalist, one of her favorite stories, "Environmental Economics: Are We Being Served?" (Anon. 2005), which she refers to in box 4.5, came about when we did a communications workshop for scientists at the Smithsonian Tropical Research Institute in Panama. In box 4.5 she describes what sets the *Economist* apart.

Science Magazine

For many scientists, getting into *Science* is equivalent to making the cover of *Rolling Stone*. But even if you don't get your study in *Science*, there is more than one way to skin a cat: you can always try for its news section. Some scientists I know have been amused and delighted when, after having their manuscript rejected, they get in through the back door with a marvelous news story in the front pages of the magazine.

At the 2002 AAAS meeting in Boston, Daniel Pauly, David Malakoff, and I were chatting in the hallway after Daniel had presented the new findings of his group about the decline of fish in the Northeast Atlantic. I playfully mentioned to David that his esteemed employer had just spurned Daniel's sub-

---

BOX 4.5
What Makes a Story for the Economist?

Natasha Loder

As a weekly magazine we have to try and sit above the cut and thrust of day-to-day news. At the end of every week, our readers want to know what was important and what these stories mean in an international context. In a world where information is growing at a seemingly breakneck pace, our readers need knowledge—not more data.

So what I look for are ideas. Sure, I'm looking for an interesting or fun news story, but the heart of most *Economist* stories is a new, or preferably unexpected, idea that can be explained. It might be something about how smoking bans have led to a rise in drunk-driving fatalities in America because smokers are driving farther to find bars where they can still smoke. Or it might be about how attempts to regulate health claims on food products might have counterintuitive results with regards to the consumption of omega-3s.

There are probably hundreds of stories that could make it happily onto the science pages alone every week. But we actually only have room for about four; choosing can be difficult and arbitrary at times. Although many stories hail from journals or scientific institutions (which put out press releases), the best are those that I track down or generate myself. I think my favorite story was a feature about environmental services and how they were being used in Panama. It had a real impact on the debate, as well as being a fascinating story. 🖱 What I like about the *Economist* is that we are not afraid of complex topics, although, just like everyone else, time is my most precious asset.

Our stories are meant to be solid and meaningful, something that counteracts what appears to be the increasing trivialization of science by some parts of the media. This is aided and abetted by a public relations industry that churns out press releases by the yard in order to whip up publicity for their publication and boost its citations (articles mentioned in the general media are far more highly cited). The respected journals such as *Nature* and *Science* and a number of others are very good about not hyping their research, but the same cannot be said of all the journals, medical centers, universities, and academic institutions out there. This is not to mention all the NGOs who are busy pumping out scary press releases, which often crank up the fears in a release in order to grab the attention in a journalist's crowded inbox.

I think everyone in the chain needs to take more responsibility for thinking about the messages they may be either unintentionally, or intentionally, giving out. Very often when scientists complain about the way that their stories were covered, they don't realize that they have completely failed to summarize the point of what it is they do. This is not about spin, this is about scientists figuring out what their research means to the many different groups in society.

BOX 4.5
Continued

From my perspective, there are two other things that scientists need to do better. One is to work hard to find better ways of bringing life and understanding to their complex ideas. Of course this is my job as well, but the more abstract the science the more work it needs to be understood. The extent to which it is possible to cover something that is very abstract can often hang on a good anecdote, a bit of color, or a startling statistic. Scientists could also be better at summarizing work by others and putting their work in context. I hate receiving a pile of PDFs in answer to a short question. Sometimes it is appropriate for me to review earlier published papers, but too often scientists feel shy about putting their work in context, which is unhelpful.

When I'm talking with a scientist, if I'm working on a noncontroversial story, the ideal conversation is one that flows naturally, like an intellectual conversation. I like to set them up with a really good question then get them going. With the more seasoned scientist I am sometimes aware of being steered to areas the scientist finds important. I actually don't mind this, I'm usually open to being told that I'm missing the point and that something else is important. The critical thing is that I don't feel that difficult questions are being evaded. This happens very rarely, and is the reason why working with scientists is so rewarding.

If I contact a company and ask it, "is anyone doing what you are doing?" invariably the CEO will respond that what his company is doing is unique. If I ask a scientist the same thing I'll immediately get more names than I can handle, including those who disagree with my interviewee. What is wonderful about working with scientists is that they are mostly concerned with the same thing I am: the pursuit of truth. It is through this shared goal that the best science journalism emerges.

—Natasha Loder is a science correspondent for the *Economist*.

mission, and he replied, "Well that was the back of the magazine, there's always the front of the magazine." The result was a three-page profile on Pauly (Malakoff 2002), complete with the color maps documenting the sharp decline in North Atlantic table fish during the last century—which was what he cared about most.

In box 4.6, *Science* reporter Erik Stokstad gives you an inside peek at how the front and the back of the magazine work quite independently—and how to let him know if you have something that would be a good news story for *Science*.

BOX 4.6

What Makes a News Story in Science?

Erik Stokstad

A former editor at *Science*—I won't mention his name—once said at one of our editorial meetings, "Remember, we're part of the entertainment industry." Partly it was to puncture the pretense that can swell up at a nearly 130-year-old publication that is, by weight, at least two-thirds an academic journal. But he also had a larger point: the stories in the news section need to have broad appeal.

That's not easy for any weekly publication, let alone one that focuses just on science. It takes a staff of twenty reporters and editors, plus correspondents all over the world. The key ingredient, as I'll discuss later, is building relationships with knowledgeable sources.

Contrary to what many scientists believe, the news section at *Science* is a stand-alone entity, with an editorial staff that is independent of the editors who handle the peer-reviewed articles and reports. This explains why it's possible for us to cover science that is published in other journals.

What kinds of stories are we looking for? New results that answer longstanding puzzles or pose new questions. Surprising or counterintuitive findings are particularly welcome, although expected or confirmatory results may merit a story if they are truly important to advancing a field. In all of these cases, we try to be as timely as possible, writing about results when they are published in the current issue of a peer-reviewed journal or presented for the first time at a conference.

Many of our feature stories summarize recent progress in a particular area, linking papers published in the last year or two and explaining the connections and implications. Others go behind the scenes to explain the political or social dynamics of a scientific effort. We will highlight interesting researchers who are doing important or controversial work, and probe the nature of their success. We also sometimes take a look at a facility or program or person who is making a big impact but is not well known outside that particular field.

The best stories often come from plugged-in scientists or policymakers with whom we've talked over the years. They're the ones who know that, say, a new government policy or court ruling is based on a weak scientific basis, or about imminent changes in funding policy, major new research initiatives, or significant problems that occur in missions or facilities.

All reporters have sources like this, and we're constantly looking to meet new ones. A lot of it is simply trawling as we go about our job: I talk to a lot of scientists for the first time when I've seen an abstract of about their research and I am evaluating whether to write a story about it. I contact even more when I'm looking for some expert insights into that work.

BOX 4.6
Continued

A couple of things that are a big help during this process: scientists who call back quickly since I'm usually working on a deadline of a few days at most. (The "out of office" e-mail reply, with a contact person who can reach you, is a helpful tip that I'll need to figure out your cell phone number.) Talking like a human being, rather than a department chair or peer-reviewer, is essential for getting a larger number of readers interested in the work. A colorful quote is always a boon to a story, but sources who speak as nontechnically as possible are essential.

Several things set good sources apart. For one, they can step back from their work and recognize what might be of interest to scientists in other fields. It's not uncommon that, when I ask someone "so what else is going on that's interesting?" they'll only think of what's happening in their own lab. A quick counterexample: I was attending a conservation biology conference a few years ago and heard what I thought was a fairly interesting talk. Afterward, I went up to the organizer and asked what he thought of the presentation. Not actually new, he said. Then he added, "but have you heard what's going on with the tiger salamander in California? Now *that's* interesting!" Turns out it was, and a feature story came out of it. 🐭

On a related note, my favorite sources know things that haven't already been publicized by press release or reported elsewhere. It's a bit frustrating to ask what someone thinks is hot and then hear about a story that was published in *Nature* or the *New York Times* last week.

Good sources also tend to be fairly selfless. They understand that I interview more people than I can quote in any particular story. A lot of really useful interviews don't make it into print, after all; but they help me understand the context, whether there are rivalries that might color the comments from various sources, and so on.

If you know of something potentially interesting, please don't assume we've already heard about it. And if your research was rejected by the manuscript editors on the tenth floor, that doesn't mean it won't intrigue a news editor on the eleventh. So let us know. Tips are always welcome, and you can find the right person to contact on this list: www.sciencemag.org/about/meet_newsstaff.dtl.

Even if we've never talked before, please don't hesitate to get in touch. E-mail is the best way, with a brief description of what's important. It could be a new paper, a small conference, or a workshop where a new topic or problem is addressed. Don't e-mail everyone at the journal simultaneously; that can lead to confusion. Also, we have terrible problems with spam, and a capricious filter, so if you don't get a reply in a day or so, please call. Talking with experts is one of the best parts of my job.

—Erik Stokstad, a staff writer at *Science*, covers environmental research.

The Bottom Line

Scientists live their stories but are often afraid to tell them. By understanding how journalists perspectives and needs differ, scientists can become better storytellers. If you want to watch a journalist lean forward, remember the magic words, "Let me tell you a story. . . ."

Chapter 5

WHAT THE CHANGING WORLD OF THE MEDIA MEANS FOR YOU

We learn by going where we have to go.
—Theodore Roethke

So far, we've talked about how to understand and communicate with journalists who work for traditional mass media such as newspapers, magazines, radio, and television. But the media landscape as we have known it is changing. The reason for this, of course, is the rise of the internet and social media in an increasingly wired world.

This shift has gutted the traditional business models of mainstream media. Newsrooms around the world are being whittled away as they undergo repeated rounds of layoffs. Meanwhile, consumers barely notice. They are increasingly finding what they want in the burgeoning world of new media. The term itself is fuzzy and continually evolving, but it generally encompasses methods of distributing and discussing information that focus on user participation. Blogs, podcasts, and video clips, as well as social networks such as Facebook and Twitter, are expanding by the day, spawning leagues of competitors, spin-offs, and hybrids. Where is this all going? Opinions abound, but the truth is that no one really knows.

A detailed exploration of the changing media landscape could fill a book much longer than this one. This chapter is primarily for the many scientists who are curious—if not yet convinced—about the potential and relevance of these tools. If you are a new media expert you can probably skip this chapter. For our purposes, we'll stick to three basic questions: why is the media changing, how is the media changing, and what does it mean for you as a scientist?

Why Is the Media Changing?

Follow the money. When you look at how thin many papers and magazines are today, it's not just because the content is missing; the advertisers who paid for the pages are disappearing too. Advertisements, much more than subscriptions, pay for print media. The wild success of Craigslist and other free online advertising services has triggered an exodus of advertising dollars, leaving the traditional media funding structure in tatters. Media outlets are looking at new ways to stay afloat online, since that's where the eyeballs are migrating. With mixed success, established outlets have begun experimenting with ideas such as online subscriptions and micropayments—charging small sums (typically less than a dollar) for access to a single article. The *Wall Street Journal* is now successfully charging for online access to any of its material, and others like the *Economist* charge for at least some of their content. Others seem to be headed in this direction.

Some believe that nonprofit business models might hold the answer, since they can circumnavigate the pressures that commercial enterprises exert on quality science coverage. Foundations are playing an important role supporting ventures such as *ProPublica*, which focuses on investigative journalism,

---

BOX 5.1
The Fate and Future of Newspapers

The demise of print newspapers and news magazines is not merely a technology-based "update" to the news industry. Print media have been the engine that drives the news. Reporters in bureaus and in the field, from China to Copenhagen, go to where the news is happening, as it is happening: investigating, reporting, challenging. New and social media largely rely on this original reporting—responding to it, opining, and, in the best cases, conducting in-depth analysis and new synthesis based upon it. But now, the engine driving the news is running out of fuel.

Historically, newspapers have relied on a three-legged business model: advertising revenues (including a heavy emphasis on classified ads), subscription fees, and newsstand

---

BOX 5.1
Continued

sales. With the growth of the internet and the highly targeted, trackable online marketing it enables, newspapers have lost ad sponsors and subscribers alike. Overall readership is down, especially of the print versions. Many younger people are still reading newspaper content, but most of them get it online.

"I teach a class at Harvard to young scientists and engineering students and I always ask my students if they read a newspaper," says Cornelia Dean, a science writer and former science editor of the *New York Times*. "I've had students who never read on paper. I have them experience reading a newspaper. (*The New York Times*.) They are surprised there are ads. They also say that they read it more shallowly, but more broadly."

In a document outlining its new strategies in the digital era, the Associated Press identified today's breaking news consumption pattern: readers hear rumors via social networks like Twitter, then search Google news for confirmation, and then head to Wikipedia (Associated Press 2009) For example, the *Los Angeles Times*, at the epicenter of the news of Michael Jackson's death, attracted less than 1 percent of the traffic for "Michael Jackson" searches. Or consider that Nielsen data shows that average users of the top thirty newspaper websites spend less than twenty minutes per month on those sites (Saba 2009), compared to more than twenty minutes a day on Facebook and YouTube (Nielsen Online 2009).

"We want access to everything, we want it now, and we want it for free," says James Surowiecki of the *New Yorker*. This is the rub. During this time of transition, people have had the best of both worlds: "all the benefits of the old, high-profit regime—intensive reporting by experts, experienced editors—and the low costs of the new world," Surowiecki says. But, he points out, "if newspapers' profits vanish, so will their products, leaving the free online media with nothing to collect, cogitate on, or riff off" (Surowiecki 2008).

As Janet Raloff of *Science News* points out:

> Our founding fathers considered news so important to our society that they wrote in constitutional protections for news gatherers. They realized that people cannot make informed decisions on what they want their government or others to do without information about those who perform good or bad deeds. News gatherers are professional scouts and watchdogs. We shine a light on dark corners. (Raloff 2008)

Most people recognize the need for reliable sources to continue the traditional roles of original reporting, vetting, and disseminating to broader audiences. Today, perhaps more than ever, society needs what newspapers have historically provided. This is sure to motivate some creative and powerful solutions.

© www.CartoonStock.com

Environmental Health News, and E&E's *Climatewire,* among others. *ClimateCentral,* based at Princeton University, offers a new foundation-funded model for science and environmental journalism where scientists work with journalists to produce climate news for print, broadcast, and online sources (Wyss 2009).

Reading the news on pulp made from dead trees may well continue to diminish, but brand names that gather and disseminate the news are likely to survive in one form or another. What forms they will take is still in flux. Providing content to e-readers such as the Kindle and the iPad may be the next big thing. Newspapers hope to rebuild their subscriptions on such devices, which allow people to customize their preferences and receive electronic news updates throughout the day. Despite all the bad news, a spirit of optimism prevails: according to a 2009 survey by the Pew Project for Excellence in Media, the majority of online journalists feel confident that a workable revenue stream will be found.

How Is the Media Changing?

The mainstream media are struggling to maintain relevance in a digital world. As we just discussed, finding the right business models to stay profitable is a big part of this. But there have been significant shifts in the way day-to-day business is done, presenting new challenges to reporters, editors, and producers. Overall, there are fewer reporters trying to cover more ground faster than ever. The following sections provide a basic overview of the trends.

Constriction

> Science is a totally disposable beat. Politics and money are not.
>
> —Andrew Revkin

Mass media corporations are cutting staff and overall costs in an attempt to maintain or increase profits. Overall, estimates suggest a loss of roughly 25 percent of all newspaper newsroom staff between 2001 and 2009 (Pew Project for Excellence in Journalism and Edmonds 2009). Since 2008, jobs in TV and radio journalism have declined (Papper 2009), and even NPR made its first organization-wide layoffs in twenty-five years and instituted furloughs (Farhi 2009). In 1988, nearly 200 newspapers had at least a weekly science page or section (Falk, Donovan, and Woods 2001). By 2005 that number had shrunk to thirty-four. Since then things have not improved. Even the *Boston Globe*'s science section was axed. If you can't sustain a science section in Boston, which has one of the highest concentrations of academics in the country, where can you?

Yet not everything is crashing—the majority of digital news outlets surveyed by the Online News Association reported that their staff levels remained steady (30 percent) or even increased (39 percent) during 2008 (Pew Project for Excellence in Journalism and Online News Association 2009).

Convergence

> We are all in the multimedia business now—I am now essentially doing television.
>
> —Cornelia Dean

In 2008, more people reported getting their national and international news "primarily" from the internet than from newspapers (Pew Research Center for the People & the Press 2008). The demands of the online audience are blurring the lines between print, radio, and television news.

Journalists are now required to be multitaskers comfortable in a multimedia world. The transition to online reporting has everyone scrambling to provide interactive content—interpretative graphics, maps, video, venues for online reader responses, and so on. The science staff of NPR does photography and video boot camps so they can learn to collect the materials they need as they report their stories. Television reporters find themselves writing for their stations' websites. Print media are hosting blogs and podcasts. Journalists and media outlets alike are experimenting, adapting, and changing their formats as they try to gauge what their audiences want.

The trends of constriction and convergence have important consequences for journalists, the way they work, and how they interact with you.

Shrinking Timelines

Journalists are working harder and faster with less time than ever to devote to each story. With fewer reporters covering more stories, their world has accelerated, creating an even broader gap between their timelines and those of scientists. The hungry internet must be fed twenty-four hours a day with constant updates and commentary. This cycle values hot and breaking stories above all. "Many feel science stories won't disappear. They might just become shorter, more superficial, or less balanced" (Raloff 2008).

Fewer Specialists

Because of constriction, specialized journalists are becoming increasingly rare while generalists are on the rise. This is partly because science and environment sections are not major moneymakers for news outlets, so their specialized journalists are being cut. It's also partly because science reporters are more experienced, which makes them more expensive. Some science reporters are moving from traditional venues to specialty online services. Those who replace them, as well as those who stay, must cover a broader range of topics. Some of the best specialists started their careers as general assignment reporters and are now coming full circle.

More Space

On the upside, some media brands have used the internet to create space for more in-depth coverage than their paper versions allow. Andy Revkin started his Dot Earth blog at the *New York Times* as a vehicle for important environmental issues that wouldn't make it into the print version of the *Times*. Revkin notes, "Each individual post does not have the same reach as the print edition, but collectively they draw an audience of several hundred thousand a month who have an interest in these issues."

The *Economist* has added an online column called Green.view that allows more coverage for interesting environmental stories than the magazine would allow. Venues such as these are excellent places for scientists to get their information out there and have their stories told, as there is a steady need for good stories. Other event-specific forums have begun to appear featuring a range of perspectives from pundits, scientists, and policymakers as well as interactive components such as live interviews and moderated audience discussions. These venues provide a timely way for scientists to engage and for others to respond. But the content producers are only one half of the equation. This brings us to the second half of the new media revolution: user participation.

The New Media

Around the year 2000, a new set of online tools began changing the nature of the internet. Before that, the internet was just another one-way broadcast mechanism, constrained by slow dial-up connections and static website architecture. This revolution, sometimes called Web 2.0, was sparked by the development of new platforms that made it fast and easy for anyone to create, share, and comment on material online. The internet is now a user-controlled, interactive network. This fast-moving new world thrives on the energy and diversity of crowds, but can also suffer from the chaos and aggression of a mob.

The best prediction anyone can make about new media is that almost anything said today will be passé tomorrow. Rather than focusing too much on specifics, table 5.1 briefly outlines the pros and cons of new media.

Here, we briefly describe some of the new media technologies that offer opportunities for scientists. None of these is guaranteed to be the best fit for you, but it's worth at least sniffing them out. Examples, links, and more information are available on the website. 🖱

Blogs

For all their pitfalls, blogs are one of the most visible ways for scientists to engage in new media. In fact, the old distinction between "amateur" blogs and "professional" news websites is quickly becoming obsolete. Some standalone blogs are breaking important stories before the traditional media, while many traditional outlets are experimenting with reporters' blogs, treated like

TABLE 5.1. Properties of New Media

Property	Benefits	Costs
Speed:	**Real-time** conversations, news as it happens	Increased **volume** of items makes it difficult to follow conversations.
Low entry cost:	Increased **participation**, democratization, diversity of opinions, and global conversations	Unlike professionals, citizen journalists are **not accountable** for what they produce. Accuracy suffers, and too often, "user-generated content" is copied—often without permission, attribution, or links to original sources.
On-demand:	Content and delivery **customized** to user's preferences	Self-selecting processes tend to limit our exposure to contradictory information. This reinforces existing worldviews lead to further fragmentation and **entrenchment**.

columns, and embedded in newspapers, magazines, and other online news outlets. While they serve to interpret and enrich the news with analysis and context, most blogs aren't original sources. Most lack editors and fact-checkers to vet them. Yet in the blogosphere, a chaotic version of peer-review thrives in the comment sections and network of posts and replies in a discussion. As radical as it is to suggest that crowds can reliably self-edit, Wikipedia demonstrates what is possible. In any case, this "free market of ideas"–type editing doesn't serve the same gatekeeping function as in academia and traditional journalism.

"The blogosphere tends to be dominated by those with a strong point of view," Andy Revkin says. And blogs tend to attract audiences that are like-minded and self-reinforcing. These groups may become more and more convinced they are right and less tolerant of others. Ironically, while new media increases connectivity, it can also breed insularity. Carl Zimmer, a respected science journalist and blogger, had this to say at a New York Academy of Sciences symposium in May 2009: "I'm sometimes pessimistic about the isolation of blogs. Online, you go and look for what you want." Nevertheless, he says, "I look at blogs as sticking my head into the future. All the info we get about science will be reorganized online. You have to go where people are going" (Zimmer 2009).

Scientists should investigate blogs because they are increasingly influential. Mainstream journalists, freelance science writers, nonprofit workers, students, and tenured research scientists are all represented in the blogosphere. They have embraced the idea that by commenting, guest posting, or creating your own blog, you can increase your profile, attract a following, and occasionally attract the attention of journalists and policymakers.

Journalists read blogs to keep abreast of what people are saying about a given topic. They have their favorites and typically use them less for finding story ideas than to keep tabs on reactions, especially to things *they* have produced. But they will go to a blog if you direct them there with a good reason. At times the sheer buzz in the blogosphere, or a video that spreads through the web like a virus, can be the topic of news stories.

In Chapter 11, we discuss more pros and cons of scientists' blogging. In box 5.2, Andy Revkin shares why he does it.

Social Networking Sites

In their simplest form, social networking services have three main components: a profile page where you post information and photos about yourself, a network of relationships that categorizes and connects your profile page to contacts (for example, friend, colleague, business associate), and a messaging system that allows you to communicate with your contacts. In the United States, the most popular of these "personal space" social networks are still Facebook and MySpace. In late 2009, Facebook accounted for an estimated 1 out of every 4 pages viewed online and drew more than 124 million unique visitors (Drake 2009; Compete, Inc. 2009).

On the surface, the "you-centeredness" of these services leads to the criticism that users are self-absorbed narcissists. But that misses the point of making connections. You might use Facebook in just one or two of the ways Jane Lubchenco does. As NOAA administrator, she uses it to post updates about meetings and testimony, announce decisions, share photos of NOAA at work, and recognize the work of staff scientists. You might also use it to keep in touch with former students and postdocs, or to stay abreast of grassroots science policy action.

Conversation and Commenting Tools

The real business of this category of tools is recommendations, reactions, and discussions. It is a wildly diverse group that includes the microblogging

BOX 5.2
Why I Blog

Andrew Revkin

For most of my career as a journalist, the media were largely a one-way conduit for information. 🖱 Walter Cronkite's sign-off, "That's the way it is," distilled this process to its essence. The nightly network news broadcasts provided a kind of intellectual comfort food, served family-style. Newspaper front pages offered the same sense of authority, literally transforming a turbulent, fast-changing world filled with shades of gray into black and white.

As the millennium turned, weaknesses in that model became more apparent. The real complexity and murkiness of world events cut against the distilled, rationalized, and often faux sense of logic and causality in such presentations. At the same time, new information technologies were, of course, eroding the foundation of the big profitable businesses that for generations were the prime purveyor of news.

Now we're all at a 24/7 information buffet, selecting ideas and voices that, more often than not, reinforce preconceptions and established tastes rather than offering challenges. (And it's magically a free buffet, to boot.) But at least a restaurant buffet has everything on display, like the assortment of stories on a traditional printed front page. Much media content now is sought with blinders, with the simple click of a shortcut, Google search, or scan of Facebook or Twitter entries streaming from like-minded friends, associates, or pundits.

In other words, we're still dining on comfort food, but each of us is creating a personalized menu.

Both models for disseminating and sharing information, imagery, and insights have merits and weakness, but also strengths. Media that thrive online will be those that sustain both the sense of an *agora*, a gathering place, for examining the core ideas of the day, while also providing focused domains for in-depth discussion on critical, more specialized topics.

The online journals called blogs, at least in their most familiar form, will long play a vital role in this new mediascape. My effort along these lines has been somewhat different than many blogs. I don't say it's better. It's just my experiment. Blogging is implicitly an evolving form of inquiry or exposition, full of trial and many errors.

I see Dot Earth as fundamentally interrogatory. It's an open exploration of evolving bodies of ideas related to smoothing the human journey in the next half century. It is not a prescription for what to do or think. It is a model of a way of thinking. Mind you, I have a viewpoint and I am a passionate advocate—for reality. My goal is to look at issues that will determine how well humanity meshes infinite aspirations with life on a finite planet, to describe what we know firmly; what we can learn through more inquiry; what is essentially unknowable on meaningful timescales; and then what society is left with when you strip away the hype.

BOX 5.2

Continued

The more I see polarized discourse, whether on population policy or energy technology, the more I think my approach, however difficult, is necessary. If I cede the debate on something like climate change to the most strident voices, that could encourage society to pull back from considering such issues in the face of what appears to be just another yelling match.

I generally try to build outward from the established base of information that is not in contention, rather than staking out a position—global warming is an unfolding catastrophe—and picking and choosing facts and anecdotes that support the conclusion while ignoring those that don't.

And that's why I sometimes get a lot of heat from, well, everybody, because that's an approach that doesn't always suit a policy agenda—particularly in a country where politics is utterly polarized and, too often, played out from the edges.

I think the *New York Times* and conventional media serve a very important purpose. I think blogs now serve a very important purpose, too. The web generally has great potential to do good and bad things—I think overall the good will prevail.

After spending a quarter century working in conventional journalism and then creating my blog, I left the *Times* to move to an academic position. But it's useful to note that the course I'm building at Pace University will be, in essence, an educational variant of my blog. I won't be teaching so much as colearning with the students in front of me, and those participating via the web. Our daily question will be the same one I've examined on my blog. It boils down to a formula: 9 billion people + 1 planet = ?

My shift was part of a natural process that is moving ever more of the activity we call communication out of the wedge called media/journalism and into the wider cloud of discourse. In the face of such changes, one reality for the community of scientists who might once have relied on conventional media to tell their story is the need to get more involved in outreach and interaction.

Just because I have chosen to keep blogging doesn't mean that it's right for anyone. But whether through a blog, or simply by becoming an engaged, involved consumer of journalism, blogs, and—yes—even Twitter, a scientist interested in how society perceives his or her field can help tamp down distortions and misinformation and keep society attuned to the value of knowledge-building.

—Andrew Revkin writes his Dot Earth blog under contract with the *New York Times* since joining Pace University as Senior Fellow for Environmental Understanding.

services like Twitter and Google Buzz, bookmarking services like Digg, and aggregators like FriendFeed. Don't get lost in the silly jargon—tweets, tumbles, pokes, whatevers. What you need to know is that all of these services capitalize on the value of social networks as personalized filters for separating the interesting from the annoying in an increasingly noisy world.

For example, Twitter asks the basic question, "What's happening?" and though some do take that literally, telling you what they are eating for lunch, the application works because you only have to listen to those you find interesting. That might include scientists sharing ideas and insights from a conference, Ira Flatow giving backstage commentary and updates about the next episode of NPR's *Science Friday*, or GenBank announcing that it has posted its thousandth complete microbial genome. The idea is to find new people who are useful to you and, if you want, actually talk about shared interests. You can tune your personal network so you get exactly the kind of news, links, and commentary you want to see, in real time.

Twitter is a tool that can be extremely useful when all hell is breaking loose and you need information fast. During the Santa Barbara Jesusita Fire in 2009, Liz Neeley, COMPASS's assistant director of science communications, was part of an NCEAS working group on salmon. The fire was moving fast in multiple directions, and evacuation orders were being issued moment by moment for various neighborhoods. Using Twitter, Liz knew was happening—as it happened, and often far ahead of the news—because she was getting updates from firefighters at the front lines and people all over the city. With that information, she was able to plan ahead, keep the rest of us informed, and avoid roadblocks and traffic snarls when it was time to get out of town. This kind of tool might prove helpful to you in the field as well.

Social bookmarking sites like Digg, Delicious, StumbleUpon, and others use tagging and ranking systems to help you find the websites, videos, and images that will be most interesting to you. These sites not only point you to material of potential interest, but they can be game-changers for articles featuring you and your work.

Digg, for example, is an amplifier. In March 2009, *Nature* journalist Erika Check Hayden did a story about leaf-cutting ants. 🐭 It had a second bounce when people who might never read *Nature* discovered it on Digg. Here's how it worked. Some readers found Erika's story and liked it. They submitted the link to Digg, along with keywords—like "science, nature, ants, antibiotics." Then other Digg users came across the story, either by browsing or if they happened to search for those keywords and similar ones. Users can vote a story up or down, depending on if they like it. If a story gets a lot of positive votes, it rises higher and higher in visibility, until like Erika's story it makes the front page of Digg—and becomes the first thing people see when they come to the site. As of this writing, Digg averaged 58 million unique monthly visitors (Arrington 2009)—that's six times more than the *Washington Post*'s website. So if you catch the fancy of Digg users, this new media outlet can potentially introduce your story, photo, or video to a gigantic new audience.

Finally, aggregators like FriendFeed can efficiently simplify and consolidate all the conversation streams and updates from across your social networking and bookmarking websites, blogs, micro-blogging updates, and so on. FriendFeed pulls all the conversations you and your contacts are participating in or following—like comment threads from different blogs, Facebook, Twitter, and so on—and shows them to you in one place. This means you only have to go to one site instead of twenty to follow what's happening, and again, you can tune your aggregator to reflect your interest in any of a range of topics, from the open access publishing movement to bioinformatics.

Multimedia Sharing

This final group of new media tools includes applications for sharing photos (Flickr, Picasa), video (YouTube, Vimeo), audio (Podomatic, iTunes), and more. You might find these sites useful for finding materials you can use in presentations or, alternately, for attracting attention to your most recent research paper, field trip, or lecture.

Crowd-sourced recommendations are as important in this realm as they are in social bookmarking networks—in fact, it's really the same thing. Some people will find your materials by searching for them, but most will run into them by following what their friends like and recommend. This means that you want to start by advertising your photos or podcast, whatever you might have, to your own friends and contacts, by announcing them on Facebook, by tweeting about them, or by other means. Given the right material, timing, and promotion, your video could even go viral—one clip of a megamouth shark has been watched more than 4 million times since 2007, while one that explains supersonic booms has been viewed more than 9 million times.

In all, new media technologies offer powerful, useful tools for sharing your work and for staying on top of the online trends and developments that are most useful to you. Visit our website for a current selection of examples, how-to advice, and links to learn more. 🖱

What Does the Changing Media World Mean for Scientists?

Sticking to traditional peer-review publications is clearly not enough in this new world of diversifying media and ever-shorter attention spans. *Nature* magazine is taking an energetic role trying to convince scientists to get involved in new media. In an editorial entitled "Filling the Void," the editors call out to scientists:

> As science journalism declines, scientists must rise up and reach out. An average citizen is unlikely to search the web for the Higgs boson or the proteasome if he or she doesn't hear about it first on, say, a cable news channel. And as mass media sheds its scientific expertise, science's mass-market presence will become harder to maintain. (*Nature* 2009)

New media can remove the intermediary and provide scientists with direct connections to new audiences. The changing world of the media makes it more important than ever to rise above the noise and be able to communicate your own messages effectively. Specialized science and environment reporters are getting rarer by the day, meaning you can't rely on them alone to get the message right on your behalf. At the same time, the internet provides access to more information than ever before, but it also enables misinformation movements such as the global warming skeptics and antivaccination lobbies. "For all the good stuff than can be put out there, there are just as many people empowered as never before to cluster around the really bad stuff. And that is precisely what we are seeing," says Chris Mooney, who writes a blog called The Intersection for *Discover Magazine* and is the coauthor of *Unscientific America* (Mooney 2009).

How can busy scientists hope to hack their way through the "new media jungle"? While younger scientists are more comfortable engaging with this sociological experiment, it's tempting for many scientists to throw up their hands and say, "I don't have time for this." But scientists take themselves out of the game when they don't engage. Carl Safina of the Blue Ocean Institute warns that "scientists are reaping the results of their own irrelevance due to their inability to communicate their science" (Safina 2009).

Railing against the changes, or even ignoring them, is an abdication of responsibility. More than ever, society needs scientists who can communicate in the new world. By cutting through the chatter and breaking down the barriers that isolate them, scientists can productively engage in society's debates.

Publicizing your science in the mainstream media often generates the biggest splash. At present, most policymakers and the public still rely on the mainstream media as gatekeepers to determine what's important. Yet social networking and new media are increasingly seeding stories and ideas that influence the mainstream. As the new ecosystem matures, and traditional media innovates, the distinctions between the two are becoming increasingly blurred.

So, your strategy should be multifaceted. Consider exploring these new venues as well as the traditional ones. Try your hand at blogging, using Facebook, or posting video on YouTube. When was the last time you updated your website? At the very least consider investing effort in the design, content, and usability of your site so that people (including journalists) can find you and learn about you and your work. There are many examples of well-done research lab sites, so we've highlighted a few on our website. 🐭 Ask your colleagues what they find interesting and useful. There are many opportunities out there worth investigating.

The Bottom Line

The mass media and new media are increasingly linked. To be heard, scientists will need to be increasingly media savvy and track the ongoing evolution of new communication forms in order to stay in tune with the changing landscape. And as the editors of *Nature* point out, "scientists should encourage any and all experiments that could help scientists penetrate the news cycle" and, I would add, expand beyond their traditional audiences however they can to communicate.

Chapter 6

WHAT YOU NEED TO KNOW
ABOUT POLICYMAKERS

Half of environmental reporting is science and the other half
is politics.

—John Heilprin

What sets the cultures of science and policy apart? Scientists are collectively
driven to answer the question "How does the world work?" while policymakers have to decide, "What should we do?" This fundamental divide in focus—exploration versus action—affects the day-to-day work of each group,
including the information they seek and how they use it.

Policymakers make decisions on record for a living. They want to get
things right because they are accountable to colleagues, opponents, and, most
important, their constituents. Unfortunately, the definition of "right" is never
clear-cut for policymakers. It varies widely depending on their personal
politics, the interests of their constituents and colleagues, the latest election
cycles, and current events. The pressure-cooker world of policymaking is
largely disconnected from the scientific community, sometimes producing
policy choices that don't account for established scientific knowledge.

This leaves scientists frustrated. Thus, understanding the policymakers'
culture and their needs can open doors and help scientists gain a seat at the
policymaking table.

Policymakers consider the five Ps: people, press, policy, principle, and politics. But for all of the diverse decisions that they must make, their work applies three fundamental questions:

- What is the problem and do I have responsibility and authority for it?
- Who will it affect? Are they my constituents?
- If I take a specific action (or don't), who wins and who loses?

These questions open the door for scientists to get involved. Many researchers feel most comfortable with the first step: defining the problem. Policymakers cannot respond to a problem they don't see or clearly understand. Science is good at characterizing changes to natural systems, providing early warning of unnoticed or potential problems, and helping define their scope and scale. Science can also play a critical role in helping to identify who will be affected in given scenarios, and how.

Policymakers *want* to hear more from you. But to be successful, you have to reconfigure your approach once again. Scientists are excellent at asking questions, challenging assumptions, and providing plausible alternatives. You deal with methodological considerations and incremental progress. In contrast, policymakers look to science for clear, unambiguous answers—even if that answer is "our educated guess." (More on this in chapter 7.)

Cultural Comparisons

Scientists and policymakers share some personality traits too. Both tend to be competitive and analytical with a round-the-clock appetite for their work. Contrary to what many scientists think, policymakers' jobs can be every bit as intellectually demanding and time-intensive as their own. Both take pride in their work and strive to advance their field. But their modus operandi and work environments are at odds.

Policymakers are generalists. They are more like journalists than scientists in that they need to be quick studies. The U.S. Congress, for example, grapples with far more than we see reported on the nightly news. More than 10,000 bills are introduced in each Congress, and a legislator can expect to cast thousands of votes on as many as 500 bills per year. All of that typically happens in less than 150 working days. The setting is fast-paced, with members and their staff juggling hearings, floor votes, debates, office meetings, policy negotiations, press, constituent advocacy, and much more. It is not possible for a single person to develop a deep understanding of each issue.

BOX 6.1

A Day in the Life . . .

Here is Senator Max Baucus's schedule for a single day. And these are only the activities he actually had on his calendar; he also had scores of walking meetings and quick consultations with staff and other senators. Baucus and his staff are constantly adjusting his priorities to make sure he's getting the most out of every minute of the day.

March 25, 2009

- 8:15 A.M.—Montana Constituents' Breakfast with Senator Tester
- 9:00 A.M.—Meeting with Senior Staff
- 9:30 A.M.—Meeting with Depart of Transportation Secretary Ray LaHood
- 10:00 A.M.—Environment and Public Works Committee hearing entitled "The Need for Transportation Investment"
- 10:45 A.M.—Press Interviews
- 11:20 A.M.—Speak at Baker Hostetler's 20th Annual Legislative and U.S. Government Policy Seminar
- 11:45 A.M.—Meeting with Brad Garpestad of Great Falls, MT
- 12:00 P.M.—Vote on Senate Floor
- 12:30 P.M.—Senators-Only Luncheon Caucus to discuss the budget with President Obama
- 2:30 P.M.—Meeting with students from Troy High School, Broadwater High, Libby High, Harrison High, and Senator Tester
- 3:00 P.M.—Meeting with Carol Lewis
- 3:30 P.M.–4:00 P.M.—Meeting with Senators Boxer, Inhofe, and Voinovich
- 5:30 P.M.—Meeting with Senators Grassley, Rockefeller, Hatch, Dodd, and Gregg, and White House Office of Health Reform Director Nancy-Ann DeParle regarding health care reform
- 7:00 P.M.—Dinner with White House Deputy Chief of Staff Jim Messina

—Taken from Senator Baucus of Montana's website: http://baucus.senate.gov/Daily%20Schedule.html.

Scientists' research interests are divided by natural boundaries that are fuzzy and flexible—species, watersheds, and ecosystems, for example. On the other hand, policymakers' interests are bounded by society—such as borders between states or electoral districts, or jurisdictional lines between federal agencies.

Scientists often need big blocks of time, free of meetings and calls, to get their most important work done. For policymakers, though, meetings and conversations *are* much of their work. Each day, every congressional office has hundreds of contacts, including phone calls, e-mails, and in-person meetings. Because policymakers have to balance constantly shifting priorities, keeping a running dialogue open actually makes their work possible.

Timing is another mismatch between scientists and policymakers. While Congress may grapple with particular issues continuously over years or even decades, the actual decision points can happen quickly. Once a policy is finalized—for example, when a law is passed—that issue is effectively closed for several years as Congress turns its attention to other pressing issues. Legislators can't wait two years for the result of a scientific study before they make up their minds about a vote they have to cast next week. By the same token, important research results might languish for years, or never be considered in the decision-making process at all, if the timing is wrong and no one actively works to bring them forward.

Policymakers in high-level bodies hardly suffer from a lack of information—they have too much. But they do need effective filters. Information

comes from many sources, most of which have explicit political agendas. For example, when a legislator must decide how to vote on a bill, the players who may try to influence him or her include the executive branch, the states, industry lobbyists, nongovernment organizations and their lobbyists, and, above all, constituents. Each party supports its case with evidence that sometimes includes science. When a groundbreaking new study with direct policy implications gains attention, it's not unusual for opposing groups to read very different meanings into the same results. Policymakers need help cutting through this noise. This is why they are not interested in the details of your science—they want to know what it *means* for the decisions they must make.

Local-level policymakers, such as county and sometimes state or provincial government officials, may not suffer as much information overload. But they still need help finding the right information and understanding what scientific findings mean. In contrast, legislators in some smaller states and local governments are often hungry for scientific information. If you make yourself available, you can become a trusted source and help them see how and why science is relevant. It is often easiest to get your foot in the door at the local level, so this can be a good place to start.

Common Misperceptions

In conversations with scientists and politicians, the following themes resurface time and again.

Scientists' Perceptions of the Policy World

My Efforts Won't Make a Difference

Scientists commonly assume that their contact with a policymaker will be treated with polite indifference at best, and complete dismissal at worst. While this is sometimes true, policymakers do pay attention to relevant scientific expertise. You can better your chances of being heard by knowing your audience. For example, you won't get very far by talking to a member from landlocked Oklahoma about the effect of coastal armoring on beach formation.

However, policymakers are likely to welcome your help on a specific issue they're actively considering. In a meeting with those crafting a bill to protect coral reefs, Nancy Knowlton, the Smithsonian Institution's Sant Chair for Marine Science, was asked to comment on whether the bill language would achieve particular ecological goals. In reply, Knowlton described the term "resilience" as it applies to coral reefs and the role it could play in management. The term now permeates the bill. It took about a year for the language to evolve, but Knowlton's conversations directly affected the final product.

I Only Met with a Staffer—I Wasted My Time

Scientists feel insulted when they take the time to meet with a policymaker only to find their meeting interrupted, or are asked to "walk and talk" through noisy halls instead of sitting down for a personal meeting. Worse, they may be given over to a "twenty-something" staffer. If this happens to you, don't get upset—these situations are routine. In fact, staffers play a critical role and have enormous influence.

A good staffer acts as a gatekeeper and steward of information for the member, providing much needed expertise on a specific set of issues and helping them "sort the wheat from the chaff." Each office employs several staffers who serve as experts on different topics such as logging, defense spending, and health care. Even if you get a member's ear—and you shouldn't pass up any opportunity to do so—it is critical to connect with a staffer if you really want your information to be remembered and used. Make the most of any opportunity to get your message across and offer to be of further help. Establishing a good connection with a staffer is crucial if you want to be a continuing resource.

They Don't Understand Science

Policymakers are often accused of ignoring science, not understanding it, or not caring enough to put in the effort required to grasp it. It's important to realize that Congress is not a scientific institution. While most members of Congress have more formal education than the average American, not many of them are trained as natural scientists. Many are lawyers and some are social scientists—and political scientists, at that.

But there is another, bigger, issue at play here. Even when politicians do understand science and its implications, balancing the other factors—including economic, social, cultural, and political considerations—can lead to a decision that appears out of touch with scientific evidence. For example, cars produce much of the air pollution that is known to cause death from heart and lung ailments. But no one would dare propose to outlaw cars—the economic and cultural costs would prove too disruptive. Keep in mind that nearly every citizen puts the highest priority on economic prosperity and public safety. There is an important difference between not understanding science and not basing decisions on science alone.

It's Who You Know, Not What You Know

While it is true that policymakers do not turn to annotated libraries of peer-reviewed articles for answers, that doesn't mean they don't gather scientific information. In some sense, policymakers perform a sort of cloud computing: rather than personally accumulating all the information, background, and expertise that any given problem may require, they draw upon a network of trusted, remote experts. For policymakers, relationships are everything—who you know *is* what you know.

They Never Get Anything Done

We all have seen sensational headlines about corrupt politicians who accomplish little of substance yet live a life of luxury. While there are some bad apples, many policymakers are incredibly hardworking people who dedicate themselves to fighting for their constituents' best interests. Their travel schedules are grueling, and most work many more weekends than they take off. They rarely have a private moment.

The seemingly glacial pace of movement in bodies such as the U.S. Congress is not always the result of inaction. Sometimes a thorough examination of an issue leads to the conclusion that nothing more should be done. Because legislators have to address such a great number of issues in reasonable depth, even slow movement on all fronts requires an enormous amount of work.

Only Democrats Respect Science

While each party does have defining tendencies, political parties are not monolithic. Generally, a member of Congress's constituents will exert a stronger influence than a member's party affiliation on his or her relationship to science. For example, former Congressman Sherwood Boehlert, a Republican from upstate New York, served as the chairman of the House Committee on Science for many years. At its zenith, he had more than a dozen PhDs on his committee staff—the highest concentration of scientists on Capitol

Hill. He took his staff's counsel very seriously and routinely chided his peers on both sides of the aisle for "being for science-based policies until the answer wasn't what they hoped. Then they want to go to Plan B."

Politicians' Perceptions of Scientists

As with scientists and journalists, the misperceptions go both ways. Here are some common views that many decision makers hold of scientists.

You Represents Your Entire Field

In the policy world, scientific consensus—particularly on contentious issues like climate change—is both incredibly important yet very difficult to come by. This presents a particular challenge when policymakers try to find spokespeople from the science community. Politicians generally do not invite scientists to speak about their specific, personal expertise on a narrow subject. They want to hear the reactions and opinions of someone who can speak for their field.

Occasionally, the definition of *field* is stretched to its limit. Chad English, the COMPASS director of science policy, describes a brush with this phenomenon as a staffer on Capitol Hill:

> I showed up as a starry-eyed oceanographer, ready to help solve the nation's ocean management and science policy problems. I was ready for anything . . . except an assignment from my chief of staff: "Go sort out this controversy with salvage logging." I was being asked to evaluate a raging scientific dust-up in a field that I had never studied [because I was a "scientist"]. I quickly found out that this was not an unusual experience on Capitol Hill.

Expect to be asked questions outside your specific subdiscipline. Think about questions related to your work, and think about how to answer them. When you walk into the room with a policymaker, approach them as an ambassador for your field. Above all, know your limits—be prepared to answer questions to the degree you can, and refer them to others when you can't.

You Will Give Them Anything Except an Answer

Science is about process. But policymakers need products—actionable information—for decisions that have to be made *now*. Scientists feel comfortable discussing uncertainty and unanswered questions. Policymakers complain that if you strip a scientist of conditional statements, caveats, citations, and calls for further study, there is very little left other than indecipherable jargon.

Politicians want to hear a bottom line, even if it is an opinion. For example, if you were confronting a problem with limited resources as a scientist, what first step would you take to solve it? This helps a policymaker who is trying to decide where to focus available resources to place the problem in context. It helps to compare what actions would bring results against what is and isn't feasible. Always be ready for the question, "What should Congress do?"

You Are Just Another Interest Group

Everyone who comes through a policymaker's door has an "ask," a request for that policymaker to make a particular decision or take a specific action. This informs how he or she will view you. In a policy context, the ubiquitous call for more research can be interpreted as a self-serving, thinly veiled ploy for more money. From a politician's perspective, calls for additional funding are especially egregious when a scientist is hesitant or apparently unable to give concrete answers to existing policy questions.

Furthermore, policymakers sometimes mistakenly believe that scientists are self-reinforcing and publish in order to support one another's findings, rather than critique each other. Policymakers find it annoying when they ask a scientist a question and the response is, "Well, we need to do more research." Unless you are specifically speaking to a policymaker about funding issues or are asked directly to do so, don't call for more research money, as needed as it might be. It's often more effective to focus on your issue and let someone who specializes in science funding issues make that point.

The Bottom Line

Scientists can contribute to science-based policies by learning more about the political world—accommodating its needs and tolerating its quirks. By making yourself accessible and answering policymakers' questions as clearly and directly as possible, you'll gain entrée to their world.

Chapter 7

INFORM MY DECISION:
WHAT POLICYMAKERS WANT FROM YOU

Congress is not interested in science, per se, but in the social
goods it can provide.
　　—Michael Rodemeyer

Policymakers are bombarded with "information" from those wanting to influence them. As President Lyndon B. Johnson said, "Doing what's right is not
the problem, it's knowing what's right." To sort out the nuance of complex issues, decision makers need information sources they can trust. Many scientists assume that they don't have much to add or that their input can't possibly
stand up to competing interests. Yet scientists can play a vital role, and policymakers frequently bemoan the fact that far too few make themselves
available.

"It's critical that scientists become involved in the policy process," says
Jessica Hamilton, the natural resources policy adviser for Oregon Governor
Ted Kulongoski:

> There are a lot more opportunities for agencies and government to use sci-
> ence more than they currently are doing. We need scientists to identify what
> the problems or threats might be for a particular situation. It's important
> that academia engage with agencies to ensure that agencies are using the
> most recent science.

Where Do Policymakers Get Scientific Information?

The short answer is, not from the same sources you do. With few exceptions, policymakers do not read scientific papers or attend conferences. Practically nothing presented in traditional scientific formats can easily fit into the policy process. The way you have been trained to communicate leaves policymakers out of the loop.

News

Policymakers care about what matters to their constituents, and one key way to keep a finger on their pulse is by paying close attention to the news. Typically, the media sets the day-to-day agenda. If journalists are paying attention, politicians will too. News saturates Capitol Hill: go into any congressional office and on every desk, sitting next to the computer, you will see a television or two. Staffers rely on cable channels like CNN, Fox News, and MSNBC to monitor the daily flow of news. Newspapers and news magazines will be scattered around the room, particularly the *Economist*, the *New York Times*, the *Washington Post*, and the relevant regional papers like the *Chicago Tribune*, the *Boston Globe*, and the *Los Angeles Times*. Every staffer's inbox will be packed with daily news updates from policy outlets like E&E News, *Congressional Quarterly*, *Roll Call*, *The Hill*, and *Politico*. And finally, staffers will be receiving highly specialized e-newsletters like SpaceNews. Although particular media choices may vary from office to office, the news is the one source of information that every office consumes in common.

On top of all that, they'll also be receiving blog alerts and content from new media sources. Members are increasingly using social media, like Twitter and Facebook, to be more connected to their constituents. Some, particularly party leaders, are using it to push their agenda. As these new forms of media and information sharing become integrated into political campaigns, you can expect them to play a more prominent role in the day-to-day business of Congress and the White House.

Reports

Although the media are ubiquitous and influential, there are other, more formal sources of information available to Capitol Hill. The National Academies respond to commissions by Congress and the federal agencies to provide authoritative and detailed reports on a huge array of topics. The Congressional Research Service (CRS) is an arm of the Library of Congress devoted to providing Congress with research and analysis on legislatively relevant issues. CRS reports are exhaustive and often include policy-relevant analysis of related science

Government agencies such as the Environmental Protection Agency, the National Oceanic and Atmospheric Administration, and the U.S. Geological Survey also produce reports and white papers to satisfy legislative demands and to advance the White House's agenda—which may simply be to share the results of programs initiated by the administration. Many agencies have science advisory boards that focus on the science relevant to the agency's mission and jurisdiction. Independent commissions chartered by Congress, such as the 9-11 Commission, provide rapid, focused analyses from a breadth of perspectives and generally include substantive policy recommendations that cut across committee jurisdictions.

Hearings

Hearings are formal, official congressional events called by committees and subcommittees. They feature a number of speakers who deliver prepared remarks directly to the members of the committee, typically for five minutes each, and then answer questions.

Hearings vary in length from a couple of hours to several days, and attendance can range from a handful of people to hundreds. One or two members of Congress may attend or the entire committee might show up, depending on the perceived importance of the issue. Hearings provide a valuable opportunity for members to listen to a variety of perspectives on an issue and to develop an official record on which to base their lawmaking. In the policy realm, hearings are the rough equivalent of a published result. Once testimony is on the record, it can be cited and used to advance a political agenda. (For advice on how to be effective if you are asked to give testimony, see chapter 13.)

In some cases, hearings are more theater than substance. Members of Congress may want to demonstrate their concern for an issue, although they might not intend to follow up with immediate legislative action. Alternatively, they may have already decided what controversial action they plan to

"MADAM SPEAKER, THIS *ASTEROID ALARMISM* IS THE BIGGEST HOAX EVER!"

take, but want to raise the profile of an issue in the public's eye to build support in advance.

Briefings

In contrast, briefings are less formal and are targeted at staff rather than members of Congress. They are generally arranged by outside organizations and attempt to provide information, policy options, or perspectives that the sponsor feels warrants more attention.

Relationships

And finally, as we discussed in chapter 6, a policymakers' colleagues and constituents, as well as lobbyists, NGOs, and other advocacy groups, are always looking to share information and offer putatively science-based arguments to advance their agendas.

It's critical to understand—and be able to speak to—the context in which policymakers think. If you can speak to policy and management dis-

---

BOX 7.1
A Scientist on The Hill: My Introduction to the Policy World

Chad English

I arrived on Capitol Hill, four days after having printed out and submitted my PhD dissertation, for the year-long John A. Knauss Marine Policy Fellowship. I expected to spend my time helping senators and their staff sift through the volumes of new knowledge being produced by the science community and turn it into robust policy. I thought I would be reading journal articles, explaining them to the senators and my colleagues, and writing up summaries and recommendations based on them. I was taken aback when I asked about subscriptions to any science article databases. We didn't have any. So I was ready to make do with several individual journals. Except that the only subscription we had was to the journal *Science*. Until that moment, I had assumed that any science that was of enough importance would surely be delivered to the desks of all relevant staff and members of Congress. I was wrong.

I spent two years working on Capitol Hill as staff for the two congressional committees that have "Science" in their names. In that time, I only read a total of two journal articles in the course of my work (five if you let me count the three papers where I read the abstract and conclusions and flipped through the figures). It simply wasn't how their world works.

Don't be too quick to judge Congress for its lack of scientific understanding. All the journal access in the world won't make a difference. Decision makers need information on an "as needed" basis. You can contribute by making yourself a resource they can use.

—Chad English is the COMPASS director of science policy outreach.

---

cussions that are already under way and understand how you fit into a policymaker's information-gathering network, it will make what you are sharing even more relevant.

Why You?

In chapter 2 we explored some of the different roles scientists might adopt in engaging with society. Whichever role you feel best suits you, consider that if you are not on Capitol Hill talking to Congress about your science, then it is

(a) being talked about by somebody else, most likely someone who is not a scientist and has an agenda; or (b) not being talked about at all.

"Research needs to be regularly shared with policymakers because we don't have time to go searching for it," says Jessica Hamilton.

> This means thinking about different channels. A front-page article might get our attention, but it often takes an individual to talk to us about *why* that problem is an issue. So, scientists need to make decision makers aware of what's going on, and when it's important enough, raise the red flag.

To decrease the chance of being labeled an advocate, Hamilton suggests that you "Deliver the same information to everyone—Democrats, Republicans, conservation groups, everyone. If you are saying the same thing, it's less of a risk."

Lori Sonken, the former democratic senior policy adviser to the House Committee on Natural Resources, advises:

> Scientists don't like to lobby and they are not supposed to lobby, but they can educate. Even if they take a position on a bill, that's okay as long as they explain why. They can come in and say, "I'm coming to you as a scientist, and I'm not trying to lobby. This would be my position on this issue for this reason."

Policymakers find it frustrating when scientists or scientific associations approach them but don't really have anything to say. Representatives of a scientific association came to see Sonken when she was seeking advice on a bill that would radically rewrite endangered species law. "They were scared to give me comments on the bill because their organization had not formed a position. I was busy, the bill was about to go to the floor, and they didn't have anything to contribute. They came in the door without an ask, and without a give," she says. "It was a waste of my time."

David Lodge, the director of Aquatic Conservation at Notre Dame University, has worked a lot with policymakers and offers this advice: "Policymakers don't care—or even spend time thinking about—whether they're pushing you over the line into advocacy, or out of your comfort zone. They want answers. Your professional integrity is not their direct concern." So, he counsels his students, "Define your personal boundaries and be sure you've thought through how far you're comfortable going *before* you step into the room with a policymaker."

Your Role as a Resource

Congress and other decision-making bodies accomplish their work through discussion. Because the debate is fragmented and evolving, policymakers don't always know what questions to ask of the science community, and scientists don't know what questions most urgently need answers. Thus even the most well-intentioned and thoughtfully executed efforts to deliver science via reports and white papers may not be all that helpful.

Here is an example: imagine that there is a forestry bill before Congress and a group of scientists has distributed a briefing paper. In it, they explain that protecting buffer zones around streams improves water quality and correlates with increased fish populations. But what if the bill is subsequently expanded to include "salvage logging," where lumber is harvested in the wake of a fire? If a fire burns right to the edge of a river or stream, what happens to water quality? Do buffer zones make it better or worse? On top of these evolving questions, new special interests are being drawn into the discussion, each with a set of concerns and thoughts on how to proceed. In light of the new issues, and without further input from the science community, members of Congress will be weighing an increasing number of concerns against a single, static briefing paper that no longer directly addresses their questions.

Since the original presentation of science by a report becomes less relevant as the dialogue evolves, this leaves policymakers to interpret the science in a new context and draw their own conclusions. Without scientists at the table who are able to answer questions, attention to science dwindles as more stakeholder voices join the debate.

In short, even when the effort to offer scientific information in a static form is well intentioned and thoughtfully executed, it is not sufficient to ensure that the science effectively informs the final policy. Just as most people will talk to a lawyer to help them interpret legal language, policymakers prefer to *talk* to scientists to help them understand science. So when they have three different people telling them that the science says different things, they

will turn to people they trust to help them understand. With some fore-
thought and by making yourself available, you can become one of the these
trusted resources—a "go-to scientist."

A Go-To Scientist

Jessica Hamilton explains, "When I think of go-to scientists, they are trusted
to provide facts. They are not political: they are approachable, not conde-
scending, are straightforward, and understand if they are talking to different
groups, that they have different levels of knowledge and experience."

Go-to scientists are invaluable during emergencies. For example, in the
spring of 2007, populations of delta smelt in California's Sacramento–San
Joaquin River Delta reached such low levels that federal courts ordered a
near-shutdown of state and federal pumps. This resulted in a water crisis for
Southern California and Central Valley agriculture. As chairwoman of the
Water and Power Subcommittee in the U.S. House of Representatives, Congresswoman Grace Napolitano called for an emergency field hearing to respond to the crisis. At that hearing, Peter Moyle of the University of California Davis played a key role in helping Chairwoman Napolitano and her staff
understand how the pumps may have impacted delta smelt within the context of a complex ecosystem influenced by multiple natural and human-
caused factors.

"Dr. Moyle became Chairwoman Napolitano and her staff's go-to scientist because he had made himself available as an ongoing resource, was trusted
for his expertise, and could communicate answers to decision makers' questions clearly and effectively," says Emily Knight, who served as science adviser
to the Water and Power Subcommittee at the time. "He not only helped us
plan the content of the hearing, we also asked him to testify."

How to Make Your Science Useful

Policymakers repeatedly identify the same characteristics of helpful scientists.
Here is a brief checklist:

- *Timely*—Policymakers mostly want information that is relevant to current policy discussions. What are they debating right now? What do

they have to vote on this month? If your work is outside the current discussions, do not be discouraged. If you have some insight that you feel is important to their district that will truly impact their constituents (negatively or positively), they'll want to know about it. Finally, even though most congressional attention is focused on the here and now, there are leaders who are always looking to stay ahead of emerging or impending environmental and societal issues. Your new or recently published results are worth capitalizing on.

- *Responsive*—Much like journalists, policymakers need you to respond quickly. They know you are busy, but their timelines are much shorter than the academic world's. "Timing is absolutely critical to us," says Amber Mace, assistant secretary for coastal matters and executive director of the California Ocean Protection Council. "We are lucky if we get a week or two turnaround time in our work. If scientists are slow to respond or not accessible and we can't get them on the phone or via e-mail, they just drop down our list. Agreeing to and meeting deadlines is really essential for us."

- *Clear*—Policymakers appreciate scientists who have taken the time to distill their messages so they can make their points simply and accurately and explain why it matters. Sonken says, "Scientists know a million times more than I ever will on the topic they are working on. But they don't know how to talk to Capital Hill staff. There are some really great people out there though. Ken Caldeira, for example, is a brilliant guy but he can come in and talk simply about what a coral is and why it matters." (You can find Caldeira's testimony on our website.) 🖱

- *Actionable*—Above all, setting policy means making decisions about what to do. If you are presenting data about a problem, your goal is for policymakers to agree that something must to be done. Think about what that "something" might be, and prepare to explain what the consequences of various choices are likely to be. If you understand the context and speak the policymakers' language, you can help inform legislation directly. Sonken describes how David Lodge was helpful when they were holding hearings on invasive species:

> He actually said what he thought the law should say. And I thought
> it made a lot of sense. He understood how the process worked, so
> we followed up with him. Sometimes people come in and make

recommendations, but they don't have any idea about what is feasible, what is likely, how the policy process works. But he did. So, we'd send him legislative drafts, we'd send him language for his feedback. He was trustworthy too; that is important.

- *Local*—To the extent you can, tie everything back to a policymaker's home district. After all, those are the people he or she is elected to represent. Some policymakers may be interested in global-scale issues, but all of them will want to know about a finding if it affects their constituents. So, for example, if you want to discuss sea-level rise, Maryland legislators will want to know how it will affect low-lying areas around the Chesapeake Bay, while California legislators will be more interested in how vulnerable oceanfront homes along their coastline are to storm surges. "When scientists can draw on their knowledge of methods that may have been used in different places or times, that can be really good in presenting and contrasting science. Telling us what has or has not worked in other places, other states, that's helpful too," Hamilton says.
- *Confident*—"One of the challenges is that scientists know so much more than policymakers, but they often present their responses as though they don't know very much," Mace says. "It's okay to represent uncertainty; policymakers can deal with uncertainty better than they are given credit for."

Remember, what policymakers need most is to hear what you *do* know. Don't talk in presentation mode; instead think of it as a conversation. "The

---

BOX 7.2

An Interview with Representative Grace Napolitano (D-CA), Chairwoman of the House Subcommittee on Water and Power

When you sit down to meet with a scientist on a particular issue, what in general are you hoping to get from him or her?

We do not hear from the scientific community nearly enough, so many decisions are not made on fact because scientists keep it all to themselves. But the rest of us are ignorant of what's true. It's important that scientists talk in a way we can understand. We do it too. In DC, lots of folks talk lingo to one another. We talk in

BOX 7.2

Continued

acronyms or laws and expect people to know what we're talking about. Same with the scientific community, but when they talk to us they should treat us like children. Assume nothing.

Give me fact-based answers to give to other folks who say, for example, "There is no global warming!" Maybe because I have traveled . . . I went to South America and the Amazon River and saw that half a degree of warming waters were killing fishing villages because they did not have any fish anymore.

What do scientists do that undermines or diminishes your ability to take action on their information?

You know I don't mind stopping people and saying, "Would you explain that?" But again, they need to be able to interpret what we need from them and put it succinctly instead of going all the way around and give all the parameters, which don't mean anything to me because I am not a scientist. They need to directly answer the questions we ask them. Sometimes I say, "Please get to the point." It takes them five minutes just to do an intro, but they need to get to what we need. And in a hearing, if you use four minutes to introduce it, you only have a minute left to get to what's important. Just start when the clock starts and get into the meat of the issue.

Are there any misconceptions you wish you could rid scientists of?

Sometimes reputations are well earned, maybe they are afraid to approach us because of our title, our position, but I find they never really learned how to talk to a politician. There are politicians and then there are politicians. If they can't talk to one, then they need to knock on another door. They need to find an ear that they can pass on the information so it can be utilized.

So in addition to being able to put it in simple terms, a scientist who also has an awareness of the policy context helps?

Yes. If you tell me it's an indicator of what's to come, then I need to know that. It's not just about the life of the fish that's endangered. If it speaks to the larger picture, then I need to know that so I can make that point in statements or speeches or when talking to colleagues.

Finally, is there any advice or words of wisdom that you'd like to impart to scientists who are contemplating becoming more engaged in the policymaking process?

I would be delighted if people would contact us more. Scientists should be willing to

---

BOX 7.2

Continued

stick their necks out and be involved if there is enough validity behind their research. Also, scientists shouldn't communicate with only Washington, DC, but
also their county and state because they have a stake in this too.

—Congresswoman Grace Napolitano represents California's 38th District and is a
member of the House Committee on Natural Resources. She chairs the Subcommittee on Water and Power, which handles legislation related to western water rights and public power.

—Emily Knight, California science policy coordinator for COMPASS, interviewed
Congresswoman Napolitano on June 17, 2009. Knight was a former science adviser to the Water and Power Subcommittee.

---

discussion has to be a give-and-take so the science is effectively communicated," she explains. "It's important for the scientist to first get a feeling for
the knowledge base of the person they are talking to before launching in. And
it doesn't hurt for the scientist to say that they want the decision-maker to ask
questions, no matter how basic."

The Bottom Line

The strongest message we hear from policymakers is that they want to hear
from more scientists—but above all, they want scientists to deliver what they
need by *talking* to them and *answering* their questions. "There are some scientists who are really good at crossing over between the spheres," says Amber
Mace. "We could use more."

PART III

The How-To Toolkit

Chapter 8

DELIVER A CLEAR MESSAGE

You must unlearn what you have learned.

—Yoda

Scientists tend to think that *if only* they could communicate what they know to the public, then the issues would be resolved. As if solving problems is simply a matter of filling the empty heads of the uninformed public.

There is an entire literature in science communications criticizing this "deficit model of the public" that assumes "public deficiency, but scientific sufficiency" (Miller 2001). It's a top-down process in which scientists fill the knowledge vacuum in the scientifically illiterate general public as they see fit. Kevin Finneran, editor in chief of *Issues in Science and Technology*, puts it this way: "Scientists are sometimes like American tourists; [we] think if we just speak English loud enough, people will understand us" (Finneran 2009).

Typically, scientists see themselves as a "sage on the stage." In this mode, the sender (you, the scientist) delivers a message to the receiver (your audience) in a one-way flow, rather like a fire hose. While this classic style of transmitting knowledge works for scientists, and their students who expect to receive information in this way, it does not translate to the wider world.

A 2008 study of scientists and engineers indicates that scientists' communication tendencies are a reflection of their own values: that to know science is to love it. Many found the idea of communicating to the public, who might not share this view, "a dangerous, if not impossible task" that requires "extreme caution to prevent public audiences from misunderstanding or

Information is not enough. The public is not an empty vessel waiting to be filled, and data dumps are not the solution. To have an impact, it is necessary to make your information matter to your audience.

misusing scientific information" (Davies 2008). This fear factor contributes to science being walled off from society. In Box 8.1, professor of parasitology Patricia Conrad describes how she overcame her apprehensions.

In Davies's study, she also found discussions among some scientists who spoke positively of communicating science to the public. They described it as "highly context-dependent—and perhaps most significantly saw it as a two-way (though somewhat limited) debate." Those who had more contact with their lay audiences recognized the need for more nuanced versions of their messages. Rather than talking about a single public, they considered "publics." Moreover, the public in this case is not a knowledge vacuum but an active and opinionated body.

"People generally make up their minds by studying more subtle, less rational factors," says author Chris Mooney. "So like it or not, those seeking a broader public acceptance of science must rethink their strategies for convey-

---

BOX 8.1
Overcoming the Fear Factor

At first, Pat Conrad, a professor at UC Davis's School of Veterinary Medicine, was afraid to communicate her findings on what was causing California's sea otters to disappear. She explains, "Our life's work in research can be distorted by the media. . . . In regards to the public and policymakers, we aren't really sure what is worse, to be ignored and have the significance of our research go unrecognized or to be acknowledged but misunderstood."

To help overcome her misgivings, she did the Leopold Leadership training. There, she figured out how to communicate messages that she knew people would not want to hear: that sea otters were being killed by a parasite called *Toxoplasma gondii*, passed from cat feces to the oceans via runoff from land. The solutions include controlling feral cats, keeping cats indoors, cleaning up cat poop outside, and not flushing kitty litter down the toilet.

Conrad realized that making herself part of the story was important to win the minds and hearts of the public and policymakers. Empathy is critical to getting across her message that cats are the indirect cause of disease in sea otters and that cat-owners can do something about it. She had to make it clear that she was not a hard-hearted cat hater, she loves cats, and has three of her own. But because she knows that *Toxo* "eggs" are not killed by waste water treatment, she cleans up all the cat poop and seals it in trash bags for deposit in a sanitary landfill.

Conrad's decision to go public led to awareness in coastal communities, as well as regional and state water quality boards. Ultimately, it resulted in legislation that mandated informative labeling on kitty litter, directed funds toward sea otter protection, and reformed runoff regulations. The Sea Otter Bill became law January 1, 2007. For more details, see www.seaotterresearch.org

Reflecting on her experience, Conrad says, "Scientists have a responsibility to translate the results of their research for the public and policymakers, both directly and by telling their stories through the media. By understanding how their perspectives differ, we can become better storytellers and scientists." But first, you have to get over the fear factor.

---

ing knowledge. Especially on divisive issues, scientists should package their research to resonate with specific segments of the public" (Nisbet and Mooney 2007).

The reason to step up to the plate is that no one knows your research better than you, but you need a new way to think about transmitting your information. The window of opportunity for engaging journalists, policymakers, and virtually everyone else is narrow. Speaking faster, louder, or covering more ideas in fewer words—hoping some portion will make it through the crack before the window slams shut—never really works. To be successful your message must be easily understood, memorable, and, most important, relevant to your audience.

The following guidelines will help you convince your audience to open the window a little wider—and to keep it open.

Step One: "So What?"

These two little words—"so what?"—are what your audience is asking. Why should I care about what you are saying? Or as Cory Dean of the *New York Times* often asks, "Why are you telling me this?" If you begin by thinking about *their* values, expectations, and interests, you can translate your information in a way that resonates instead of just dousing them with what's on your mind.

When you talk to journalists, take it a step further and consider *their* audience. A journalist who works for the *Wall Street Journal* will have a different audience in mind than, say, one who works for *National Geographic*. Juliet Eilperin of the *Washington Post* explains, "I write about things that have national and international policy implications. My immediate audience is everyone from the president on down: national policymakers in the United States all consume the *Washington Post*, so we are always thinking about them."

Every audience is different, so you have to be ready to modify the way

The "So What" Prism: Answering the question "so what?" is the starting point for getting your message across.

you express your main message. Think of it as passing your message through a prism—each audience has a different "so what?" What will be uppermost in the mind of a policymaker? What do resource managers care about? What is in it for a particular NGO? What might get a specific journalist interested? Why should the public care? It's important to customize what you say and how you say it to your particular audience.

Step Two: The Message Box

> If you want me to give you a two-hour presentation, I am ready today. If you want only a five-minute speech, it will take me two weeks to prepare.
>
> —Mark Twain

Scientists know too much and struggle to simplify it. "Once we know something, we find it hard to imagine what it was like not to know it," write Chip and Dan Heath in their useful book, *Made to Stick* (Heath and Heath 2007). "It becomes difficult for us to share our knowledge with others because we can't readily re-create our listener's state of mind." Scientists are bedeviled by

"Oh, if only it were so simple."

this "curse of too much knowledge." It compels you to ignore the big picture, to delve too deeply into the topic, and to flood your audience with too much information.

The message box is a deceptively simple tool that helps you sift through the mountain of information in your mind and focus on the few key messages that will be most salient for your audience. It helps take what you know, prioritize the most important information, and figure out how to frame and deliver it.

The message box can help you:

- Explain to nonscientists what you do
- Prepare for interviews
- Refine your thirty-second elevator speech for talking to policymakers
- Polish an abstract or cover letter for a publication
- Write an effective op-ed or press release
- Storyboard your website

The message box consists of four quadrants arrayed around a central issue. Here are the questions that should be answered in each of these sections:

Issue: In broad terms, what is the overarching issue or topic?
Problem: What is the specific problem or piece of the issue I am addressing?
So What?: Why does this matter to my audience?
Solutions: What are the potential solutions to the problem?
Benefits: What are the potential benefits of resolving this problem?

The message box is especially effective because it is nonlinear. There is no need to start with any particular component and work your way through step

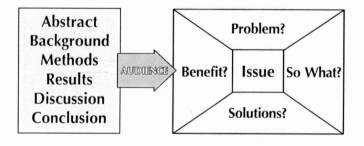

by step. If a conversation opens with a question about solutions, you can start there. The quadrant layout mentally prepares you to circle back to your main points no matter where you begin.

The message box works for any audience, but let's focus on journalists for the moment. The message box keys into the questions that journalists need to know to turn your information into a story. Used effectively, it strips away all the details that matter to scientists but can distract others and zeroes in on the core ideas that you want to convey.

As you work through the message box, keep in mind that messages:

- are the core ideas you are trying to get across, not necessarily sound bites.
- must be simple, but not necessarily simplistic. They can be explained in a sentence.
- must be limited to three or four ideas—one if it's for television.
- must be jargon-free.
- should be supported by sound bites, metaphors, statistics, and anecdotes (sparingly; see the next section).

More advice and examples are available on the website.

> A designer knows he has achieved perfection not when there is nothing left to add, but when there is nothing left to take away.
>
> —Antoine de Saint-Exupery

Despite some initial skepticism, the scientists we train inevitably say that the message box is one of the most important things they have learned. They use it time and again to prepare for an interview, to design a talk, and even to write papers and grant proposals.

Step Three: Support the Messages

Once you are satisfied that you've chiseled out the most important messages, you can add the bells and whistles. In order for your messages to hit home, you need to make them interesting and relevant by using specific examples, statistics, metaphors, and anecdotes. While your core messages shouldn't change, how you express them and make them compelling depends on your target audience.

Following are some suggestions to help you refine your messages.

The Numbers Don't Speak for Themselves, Actually

When you are immersed in your data it's easy to forget how differently the rest of the world deals with numbers. Statistics can be powerful tools for making your message memorable. But, you must use them sparingly, with an understanding of how they are likely to be received. When I tell scientists that the vast majority of journalists and policymakers can't read graphs, or figures, they find it hard to imagine. But they just don't speak that language, so whenever you present data you have to say, "What this tells you is. . . ."

In covering a story, journalists will only use one or two key numbers and, even then, probably in conversational terms. You will have to sacrifice some degree of precision, but you can make sure that your number is usable, memorable, and still faithful to the data by following these suggestions:

- *Sum it up and simplify*: Never rely on your audience to do math in their heads. Instead of reporting multiple values for different years, give the total difference. Round off the numbers and translate them into conversational terms.
- *Use frequencies instead of probabilities*: Even knowledgeable audiences struggle to understand probabilities. Studies show that values expressed as natural frequencies (for example, "only three in 10,000") significantly improve the understanding of experts and nonexperts alike (Hoffrage et al. 2000).
- *Compare and contrast*: You may have worked for months to produce a single number, but it is useless unless you tell us what it *means*. Comparisons are a natural way of putting a number into context. For example, on average, only one person dies from shark attacks in the United States each year. In comparison, lightening strikes kill almost fifty people (International Shark Attack File).
- *Explain significance versus magnitude*: Lay audiences do not understand the meaning of significance in a statistical sense. Small but significant differences can sound unimpressive and unimportant. If you can, pres-

Instead of Saying This, Try This

Instead of Saying This	*Try This*
A 90 percent increase in population.	The population almost doubled.
Annual coral cover loss was 1 percent over the last twenty years and 2 percent between 1997 and 2003.	We are now losing coral reefs more than twice as fast as we are losing rainforest.
The lifetime probability of developing liver cancer is 0.46 percent.	Out of every 1,000 people, fewer than 5 will develop liver cancer.

ent data so that the magnitude is striking. Instead of saying "we've seen a 4 percent annual decline since 1992," say "we've lost more than 260,000 square miles in the past twenty years, which is an area the size of Texas."

Think ahead: what misinterpretations are likely and how can you avoid them? How can you state your results in the most powerful terms? Sum it up, simplify, and tell us what it means . . . or someone else will do it for you and may get it wrong.

Jargon Watch

> I don't want you to pretend that you're a beat poet. I want you to sound like a scientist, but be judicious about how you use technical terms. Sprinkle a few in, like bacon bits on my salad of information.
>
> —Douglas Fox

Jargon serves a useful purpose, but only with your peers. It enables insiders to communicate concisely and precisely—to use one word instead of many. However, it is exclusive. In order to communicate your ideas broadly, the solution is the same as with any other language barrier: you need to translate.

We often hear scientists lament, "I'm just not sure how far to dumb it down." As a rule, journalists, policymakers, and the "science-interested" public do not lack the intellectual firepower to understand your work. What they lack is the highly specialized knowledge base—and the vocabulary that goes with it—that you have spent your entire professional career building. Use everyday words to get your meaning across.

Instead of Saying This, Try This

Instead of Saying This	Try This
Microbiota	Tiny living things
Hypoxic	Low on oxygen
Trophic structure	Food web
Piscivorous	Eats fish
Pelagic	Open water
Phototactic	Moves in response to light

When we talk about how most audiences lack the background to understand the language of their work, many scientists respond by asking, "But isn't this an opportunity to educate?" This is a reasonable reaction. But by putting yourself in the role of educator, you run the risk of trying to impart, explain, and define so much new information that your core message can be lost in the process. When you speak to a room of college freshmen your goal is to bring them, however incrementally, a bit closer to your level of understanding of a sprawling and complex body of knowledge. When you speak to a journalist, policymaker, or other educated nonscientist, however, your goal is to make them understand a message in the context of why it matters.

Be strategic in deciding which concepts and terms to define or explain in greater detail. There may be one or two words or phrases that are so common and so important to your work that it will actually save you time to define them. But limit yourself. Definitions you want others to learn can compete with your take-home message.

Framing Your Argument

Framing is not about spin or manipulation. As a conceptual term, "frames" are interpretative storylines that communicate what is at stake in a societal debate and why the issue matters (Gamson and Modigliani 1989).

Figuring out the most effective frame for your messages takes some thought. When Jonathan Patz of the University of Wisconsin published a review paper in *Nature* titled "Impacts of Regional Climate Change on Human Health" (Patz et al. 2005), he was determined to communicate it effectively. He used the message box to help him prepare. Below we show two versions of his message box: his initial attempt and where he eventually ended up.

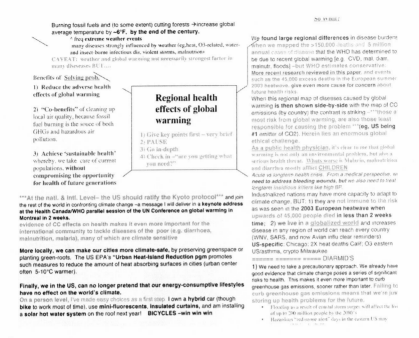

Before

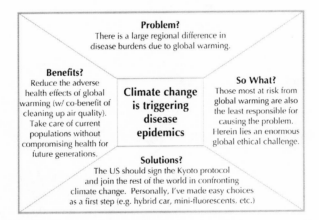

After

Like most scientists, Patz suffered the curse of too much information. When he first started talking about his paper, he sounded like the first message box appears—overwhelming. He struggled to find a clear message from the many things he wanted to say. Moreover, since it was a review paper, he had difficulty justifying the newsworthiness of his study to himself. But once he had it all listed into the message box format, he started whittling and grouping key messages. This clarified his thinking and led him to a surprise. Patz and his coauthors had not clearly articulated the most powerful message of all:"Those most at risk from global warming are least responsible for causing the problem. This is a global ethical problem." Framing his message this way expanded the relevance of their paper from an environmental health issue to a human rights issue, which he realized would have much greater salience internationally.

When he approached Juliet Eilperin at the *Washington Post* to tell her about his study, she found this message so compelling that she arranged for the *Post*'s art department to produce a graphic that wasn't included in the original paper. The *Post*'s graphic juxtaposed a map of greenhouse gas emissions by country that underscored Patz's main point. It was too late to add the second map to the *Nature* paper, but the *Washington Post* story effectively conveyed the message to the many decision makers and others who read the *Post* (Eilperin 2005).

Media the world over reported Patz's message almost exactly as he articulated it. Soon after the news broke, he was invited to give the keynote address at the World Health Organization/Health Canada meeting at the IPCC Convention of Parties venue in Montreal and many other repercussions.

Patz says, "This experience with a simple—albeit well-placed—review paper, made me realize that from one small opportunity you can have a world-reaching impact."

Making Your Messages Memorable

Try passing through a room of with a tray of items heaped in a pile. When you return, ask your colleagues (or your kids), what was on the tray? They will remember different things, and some will recall more than others. But if you leave the room and return with only four things on the tray, it is likely that everyone will remember the same four things. If you want your messages to be memorable, less is more.

When anyone starts working on a message box, it's typical to pile many thoughts in each cell of the message box. That's fine: it's part of the brainstorming process. However, the next step is to sift through these ideas and decide which is most important. This process of ongoing refinement takes time.

Once you have decided what your key messages are, the next step is to figure out how to convey them in a way that will "stick."

Framing plays a role. Patz's "global ethical issue" had traction because it evoked a moral and emotional response from people who might not care much about the environment. It's not *fair* that our emissions are making people ill in parts of the world that produce few emissions. We are wired to feel things like injustice, so this point stuck with a broad audience much more powerfully than any abstraction about emissions could have.

To really make your messages stick, you need a catchy sound bite or metaphor that will convey your complex idea in a way that is easy to understand. You will know when it works because people will repeat it. For example, "rainforests are the lungs of the planet" has been used to the point of being a cliché. It worked because it is a simple shortcut for a big idea that anyone could understand.

Steve Palumbi, director of Stanford University's Hopkins Marine Lab, says, "I've realized that most people don't live in logic-land all the time; they are feeling things. And scientists feel things too. We just tend to wrap it all up in logic and prove it." When Palumbi was writing his first general science

book, *The Evolution Explosion: How Humans Cause Rapid Evolutionary Change*, he tried to consider what people would feel about the information. To make the science relevant to readers, Palumbi worked at coming up with metaphors to describe what was going on:

> For example, you might know that taking antibiotics the wrong way creates an evolutionary arms race that leads to the drug failing to work. If you know that may be going on in your own body, if you know you could be damaging yourself by doing this, how does it make you feel?

Palumbi admits he doesn't always get it right at first, so he tests his ideas:

> Some of the metaphors I've tried are terrible. Some of them make no sense. Some of them make people laugh at you. Some of them make people look at you and say . . . what in the world are you talking about? But I keep at it until I find one that works.

Once you figure out a good way to get your core message across, there's nothing wrong with using it over and over again. Politicians, advertisers, and kids prove the effectiveness of this tactic every day.

Finally, be careful to use examples that will resonate with specific audiences. For example, in the case of the paper by Ransom Myers and Boris Worm titled "Rapid Worldwide Depletion of Predatory Fish Communities" (Myers and Worm 2003), their overarching message was that 90 percent of the big fish are gone. But if they were talking to media in the Gulf of Maine,

---

BOX 8.2

Master of Metaphor

Fisheries biologist Daniel Pauly is well known for his ability to come up with vivid metaphors that make his messages memorable. Here are a few examples:

"We are fishing for bait and headed for jellyfish."

"Long-lining has expanded globally. It is like a hole burning through paper. As the hole expands, the edge is where the fisheries concentrate until there is nowhere left to go."

"The fishing industry has acted like a terrible tenant who trashes their rental."

After the Myers and Worm paper, there was a deluge of cartoons with the headline, "90 percent of big fish are gone." Here is an example. Used with permission of Chip Bok and Creators Syndicate. All rights reserved.

they talked about cod and swordfish, and if they were talking to a journalist in Florida, they pointed to their data on grouper. They were able to get people to care about their big-picture message by using local examples.

If your message has really penetrated society, you might even see cartoons riffing off it. That's when you know you've truly made your point.

Know Your Headline

Once you have boiled all your ideas down into one message box, you are ready to distill it further until you find the one-phrase headline for your story.

I worked with Drew Harvell of Cornell University, Andrew Dobson of Princeton University and their coauthors to help them communicate a review that would be published in *Nature*, titled "Climate Warming and Disease Risks for Terrestrial and Marine Biota" (Harvell et al. 2002). While interviewing Dobson for the press release, I kept pushing him for a headline. He couldn't think of one. After we had talked for more than an hour I circled back and asked him again, "Okay, what's the headline?"

BOX 8.3
Boris Worm Discusses the Message Box

Boris Worm, assistant professor of biology at Dalhousie University, studies marine bio-diversity at the global level. Quiet and reflective by nature, Worm had a series of papers that thrust him into the media limelight when still a postdoctoral fellow in 2003. Since then, several other papers have followed. These high-profile, high-stakes papers have attracted much positive and negative attention. Throughout, Worm has worked hard at becoming an effective communicator. Early on, he learned that the message box could help. I interviewed him in October 2009.

What have you learned about getting your message across?
It's really about prioritization. You have a bunch of things you would like to talk about but really, at the core of it, if somebody has just one minute, what would you say?

What is your personal process to prepare for an interview?
I always try to be in a quiet room and to have a few minutes to concentrate to let all the clutter in my head fade away. A botched interview is often one that was given on the fly. So I look at the paper again, and the press release, and the message box, and highlight key points that I absolutely need to make. Then I think about the person who is calling—whether it's someone with an interest in ecology or economics or whatever angle. Or maybe it's a story that's more personal; this person is likely to ask about my background, so I should think through a few points about that. Journalists really value and appreciate it when they feel you are organized and on point—not fumbling around with lots of different ways of trying to explain one thing but you didn't say clearly what the one thing is. It is so much less work to interpret what the heck you mean.

Tell me about what you are doing now with your grad students to help them be better communicators?
I force them to give one-minute statements about their thesis research because it helps them be clear about what they are actually trying to achieve. It comes up all the time. People ask, "What are you working on?" Sometimes a student might say, "I am working on lobster." That doesn't mean anything. Or they give a ten-minute detailed presentation of the method they are using, but never talk about the main question, or answer the question "so what?"

Everybody gets lost in detail so quickly and the world is so full of clutter and unnecessary information that doesn't really mean much. So it's refreshing when someone is able to give facts that matter in a way that everyone can understand. And it helps the listener or interviewer understand arguments that are very com-

---

BOX 8.3
Continued

plex but are presented in a way that can be absorbed easily. That's a learned skill that comes with practice and experience with things like the message box, or understanding how journalists build a story.

I give my students the message box when they are going to be called up by media and that's when they find it really helpful. But even just in lab meetings, or when we have a visitor and I am introducing that person to my lab, I ask my students to tell the visitor what they are working on in one minute. It's light, but I find it really helps.

He stammered a bit, but then spit it out: "A Warmer World Is a Sicker World!" Bingo—that was it. Sure, there were caveats and exceptions, but this headline crystallized the authors' key message. It caught people's attention. Journalists used it. The headline went a long way to helping the authors get their message across. And it stuck.

Prepare to Dance

The point of the message box is to help you prepare ahead of time. When used to its fullest extent, it helps you determine where you want to go, yet get there in a way that sounds natural and easy. If you do it right, no one listening to you would ever know that you were working from a blueprint.

Think of it like learning a basic dance step—say a meringue. On the first few attempts it may look and feel unnatural as you try to get the basic moves down. But with a little dedicated practice you can start adding the clever stuff—the twirls and flourishes. These are the sound bites, the statistics, the metaphors, and the little anecdotes that bring your message box to life.

Think of the journalist as your dance partner. A successful interaction depends on you both feeling like you are in rhythm. I have seen some disastrous interviews where scientists have clung too rigidly to their message. No matter where the journalist tried to go, the scientist just mercilessly repeated the same message over and over again. They refused to "dance" with their partner.

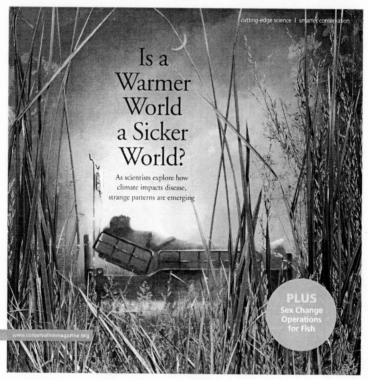

This headline has become a meme for the entire concept of climate change and disease dynamics. Try Googling it—as of this writing, some 93,900 hits come up. Posed as a question, it was *Conservation* magazine's summer 2009 cover story. © 2007–2009 Society for Conservation Biology.

In situations like these, the scientist comes off badly, and the journalist feels exasperated.

This serves nobody. You want to prepare yourself to engage in lively conversation, or address challenges, while keeping the points you want to cover in mind. Use it, and you will feel satisfied instead of panicked when the journalist says, "thank you very much—I've got a deadline, gotta run."

The Bottom Line

Less is more. Cory Dean describes an interview where she asked a scientist if he could explain something more simply. He thought about it and replied, "No." Not true! With a bit of effort, everything can be explained succinctly in universal language. The purpose of the message box is to help you do this. Many of the best scientist communicators I know swear by it—so give it a try.

Chapter 9

ACE YOUR INTERVIEW

The questions don't do the damage. Only the answers.
—Sam Donaldson

Many scientists think of an interview as an ordeal to be survived or avoided. But you can also think of it as an opportunity. There are things you want the world to know, and this is your chance. While you can never completely control how an interview will go, there is much you can do to increase the odds that you will be happy with the end result. A good interview may look spontaneous, but preparation is the key to success. Preparation gives you confidence and focus, and it shows. The odds are, you'll even find it enjoyable.

Listen carefully to journalists' questions, no matter how simplistic or abrupt, because they offer a chance to see your work from the vantage of an outsider. It's a useful reality check on how the public and even more impatient politicians might view your work.

Ingredients for a Great Interview

Don't passively answer whatever question comes—you don't want your interview to be the equivalent of robotically returning serves from a tennis ball machine, one after another. Rather, think of directing your responses towards your main points.

Like a spirited tennis match, an interview it should be a satisfying exchange between two well-matched partners. Both players should be placing their shots and making each other run to reach the ball. If it is a good interview, both the interviewer and the interviewee will have a stimulating experience. The journalist might not have much background on the topic, but journalists catch on fast. Treat the person with respect, and check in periodically to see whether he or she is with you or needs anything.

There are different types of interviews. Sometimes a journalist might call you on a "fishing expedition." They don't know exactly what they hope to find, but will know it when they see it. Offer up what you think is important and interesting.

Other times, journalists might know exactly what they want as they rush to meet their deadline with a story that is ready to be filed save for one last perspective. If this is the case, don't frustrate the journalist by digressing from the main subject. However, if you sense that he or she is open to a wider-ranging conversation, throw out a few little bits of bait to try and hook their interest in what you think is important.

Practical Tips for the Interview

Journalists quickly investigate the topic of interest and the person they want to interview. They might search for you and your organization on the web and, if time permits, quickly scan a couple of papers—but don't count on it. Most journalists simply don't have time to do any more than a cursory check. Especially if they are on a tight deadline, they will contact you and may try to get what they need there and then. If they have more time, some journalists will write out a list of questions and e-mail you.

You usually don't know how much background they have, but you want to be ready for anything. This section covers some steps you can take before, during, and after the interview to help things go your way.

Before the Interview

When journalists call you out of the blue, it's always a good idea to buy yourself a few minutes. This allows you to think about what points *you* want to convey. You might fudge a bit and say "there's someone in my office, let me call you right back" or tell them if they call you back in ten minutes they will have your full attention, but you have to mean it. Before you hang up, quickly ask for the journalist's name, number, and media outlet so you can make sure you reconnect—and so you can quickly check them out.

If you don't know the journalist, do a quick background search on Google for the reporter's name and the venue he or she is reporting for. What stories does this person or their outlet typically cover? Have they covered related stories before, and if so, what angle did they take? Do the stories appear credible, or are they "gotcha" stories? If the reporter covers a diverse beat—local politics, disasters, crime, and the occasional environment or science story—don't assume they have any background. Plan to provide some context as succinctly and respectfully as possible. However, if it's Marla Cone of *Environmental Health News* wanting to talk about contaminants, be prepared for penetrating questions. Many journalists *are* experts on their beat, and you can expect their questions to be challenging.

Collect Your Thoughts

Think about the topic. If this is something you talk about regularly, grab your message box to refresh your memory. If it's not, you might want to quickly do one (see chapter 8) or jot down a few key points you want to make. If the topic isn't fresh in your mind, look up a couple of supporting stats or specifics. Think about ways to make your points relevant and interesting such as metaphors or connections to current events. Anticipate what tough questions the reporter might ask and how you might want to answer them. Consider who else the journalist might want to consult.

Spread your message box or key points and supporting facts on your desk in front of you so you can look at them as you talk. Have a pad of paper and pen at the ready; you might write down the questions you are asked or at least a key phrase. This can help you stay on track and remind you to segue to a point you want to make.

"I think the element of my Leopold training that was most helpful was simply how to prepare in advance for interviews . . . that is, the message box," says David Conover, dean of marine and atmospheric sciences at Stony Brook University. In every interview since that training, he says:

> I knew what my message was going to be and I stuck to it. I thought in advance about what words I was going to use, the images I wanted to convey in the minds of the listeners, the analogies that could be used to explain a

complex concept in terms that a non-scientist could connect with and understand. Plus, because I had a plan, I wasn't nervous, and the aura of confidence that exudes helps assure the viewer that you really are an authority on the topic at hand. In the past, I would have just winged it and probably filled my dialogue with jargon in an attempt to sound authoritative, which instead would have come across as incomprehensible.

Most interviews that go badly are a result of not taking the time to focus first.

Natasha Loder of the *Economist* concurs: "Very often when scientists complain about the way that their stories were covered, they've not delivered any message themselves and left the journalists to think up one for themselves."

During the Interview

Remember the five Cs. You should be:

- *Concise.* Use common language, brief sentences, and so on.
- *Conversational.* A good interview is a dialogue, not a monologue.

Scientists watch Juliet Eilperin of the *Washington Post* (far left) conduct a mock interview at one of my COMPASS workshops. Photo © 2009 Ken Weiss

- *Clever.* Use examples, analogies, and visuals.
- *Correct.* Put the onus on yourself to be clear. Periodically ask the journalist if they understood what you said.
- *Cool.* Never lose it.

Illustrate your meaning with sound bites, vivid language, and metaphors. Details can be more helpful than generalities. Think about a sightseeing tour in a plane. If you only fly at 35,000 feet—that is, speaking in generalities all the time, it gets boring. Keep it interesting by zooming down to look at examples and reinforcing your points with statistics. But be careful about your use of numbers (see chapter 8). As Steve Schneider advises, "Instead of saying, '35 percent mortality,' say, 'If you were a rural farmer, three out of ten of your cows would have died.'"

As you're conducting your tour, check in to see whether you are hitting the mark and giving them what they want. Sometimes you can see the journalist's eyes glaze over. It's time to adjust your course and liven things up. If the interview is in person, or at a press conference, watch what the journalists are doing. It's similar to teaching in the classroom: if you've just made your main point and the journalists are furiously writing, that's a good sign. But if they are staring at you with their hands in their laps or are reading their Blackberries, you're in trouble.

When you are on the phone, it's harder to gauge reactions, but just go ahead and ask them, "Is this what you are looking for?" If you're off track, improve the situation while you can. Try to answer their questions more directly or make your point in a different way.

Block and Bridge

When a journalist introduces a topic you don't want to discuss, don't panic, don't get mad—you don't *have* to go down that road. Sometimes scientists tend to be overly obedient, answering any question a journalist asks you. This is the point of being prepared. You don't have to be led a like a lamb. Blocking and bridging is a useful technique that can help you redirect an interview that is heading off course. Blocking halts the direction of an interview, while bridging helps you guide the conversation back to the points *you* want to make. Together they help you regain control. Politicians frequently use blocking and bridging when they get a question that misses the point, touches on something they don't want to talk about, or is antagonistic (Kaufman 2007).

Watch them gently acknowledge the question and quickly bridge back to their main talking points.

Use this technique carefully—you don't want to antagonize the journalist by ignoring or evading what you've been asked. But you can smoothly redirect the line of questioning back to your main messages using these bridging statements:

> "I don't know that . . . what I do know. . . ."
> "The most important thing to know is. . . ."
> "The real issue here is. . . ."
> "Yes . . . and in addition to that. . . ."
> "Let me put that in perspective. . . ."
> "What you're really asking is. . . ."

These little lines can be lifesavers. Just keep your wits about you and know where you want the conversation to go.

Give Good Sound Bites

> The sound bite is your friend.
> —Cornelia Dean

Journalists rarely quote explanatory statements. The reporter can often paraphrase your explanations better than you. The sound bite, however, adds interest and puts things in perspective. It is likely to be a clever metaphor that captures your take on the issue.

You'll know a good sound bite when you hear it, like this one, by Ken Caldiera of the Carnegie Institute: "I compare CO_2 emissions to mugging little old ladies. . . . It is wrong to mug little old ladies and wrong to emit carbon dioxide to the atmosphere. The right target for both mugging little old ladies and carbon dioxide emissions is zero" (Romm 2009).

Journalists look for the human element. If you have just won a major prize, count on being asked, "How did you feel when you heard the news?" and "What will you do with the money?"

After listening to a panel of journalists, Janice Moore, a professor of biology at Colorado State University, offered this perspective:

For me it was an "aha" moment—the importance of the sound bite means that if you want to communicate clearly, it must be short, vivid, uncluttered. Are you communicating to please yourself or to get the message across? If the latter, then do what it takes, no matter how much beloved and hard-earned data gets side-lined.

Your sound bite can spark important breakthroughs. Jim Schaeffer of Trent University worked on species at risk in Ontario and wanted the government to update its Endangered Species Act. He says:

> When interviewed by a reporter from Canadian Press, I told him that the Act dated back to 1971 and that "During that time, Gordie Howe was still playing hockey, Paul McCartney had just established the group Wings, and Watergate was known only as a hotel. We need new legislation." A dozen newspapers, including editorials, picked up this quote. We made our point, and the government passed new and stronger legislation.

Journalists value scientists who give good sound bites. If they liked what you said in one story, they'll duly note it for another time. If you give a journalist a pithy quote, he or she will come back to you again.

Don't Let Your Most Memorable Quote Be Offhand

It's sometimes tempting to make a snide or cynical remark. This will likely be the most colorful, honest thing you say. The danger is that this can move your story off the mark. Journalists often seek out Andrew Rosenberg, formerly deputy director of the National Marine Fisheries Service (NMFS) and now senior vice president of science for Conservation International, because he gives great quotes, he is astute, and he can cut to the chase. But his sardonic side can derail him. "Once I was doing an extensive interview on the fight over how to rebuild the groundfish stocks in the north east Atlantic, and a reporter, at my suggestion, had done some research on fishery management in Norway as a comparison," Rosenberg recalls.

> She then asked me why the situation in Norway was different from the United States. I said, "perhaps because they don't have the attitude that everyone must get to heaven but nobody deserves to die." Of course that

was the quote in the story. Not good. It was stupid to make a cynical comment like that and therefore ruin the message.

Wrap It Up

Near the end of an interview, it's a good practice to review and clarify your main points. You can ask the journalist to reiterate what you just said, but be careful not to sound patronizing. Rather than saying, "Did you understand me?" which could imply that you doubt their intelligence, put the onus on yourself and say, "I am not sure I was clear. Do you want to tell me what you got from what I just said?" Chances are the journalist will summarize it perfectly and in a sentence or two. But it can be useful for journalists too because it helps them synthesize what they just heard. And if they didn't understand you, here's your chance to explain again and make sure they do.

Martin Krkošek used this technique in an interview with *National Geographic* television:

> The person interviewing me tried to paraphrase the research several times but got important parts wrong. I responded by saying you are correct that "such and such" *however what our findings actually mean is* . . . and would bring the conversation back to the main points we had aimed to communicate. The outcome was that the important points were reiterated several times. The interview ended up having more depth than usual, and I sounded natural, unrehearsed, and intelligent.

After the Interview

Be Helpful

If you suggested other scientists to be interviewed, be sure to send the journalist their contact information in a brief e-mail as soon as you are back at your computer. Send references you might have mentioned, but don't deluge the journalist. Send your office, mobile, or even—if you're comfortable with it—your home number and offer to be available if the journalist needs anything else. (If you do this, be responsive if he or she calls.) Journalists love it when scientists return their calls no matter what you are up to. They swap stories about scientists who called them back from an airboat in the Everglades or skiing up a mountain. This puts you on their A list.

BOX 9.1
What Makes an Interview Magic

Thomas Hayden

An interview between a journalist and a scientist is like a conversation. Or, no, maybe it's more like a dance, but with each partner trying to lead. Though honestly, there are times when it feels a little like a bad blind date, or a high-pressure job interview: as awkward as the former, with the failure stakes of the latter. Sometimes—when you get a call out of the blue about a subject way outside your field—an interview might even feel like a phone prank.

So why then are interviews my favorite part of being a science journalist? It has something to do, I think, with what usually brings journalists and scientists together in the first place. Intellectual curiosity is what gets us out of bed in the morning, and what keeps us in the lab or typing at the keyboard late into the night. Journalists and scientists have different tools of course, different needs and limitations, but we all want to know more, to understand better, to explore good ideas, demolish bad ones, and get as close to the truth as we can. And that's why the very best interviews, if they resemble anything at all, feel to me an awful lot like the best kind of graduate seminar.

If you're reading this book, you know that you should have a clear sense of what you want to say before you start an interview. That's true; that's good. But it's just the beginning. Your talking points are the short introductory lecture you give at the start of a seminar session, the key points you make to frame the discussion that follows. You want to get your main points across, certainly. But you don't want to be so worried about delivering that message that you miss the chance to go deeper.

If you've got good students, they'll have done the reading; if you've got a journalist who isn't on such a short deadline that there's only time for the bullet points, he or she will have done the prep, too. Good students and good journalists alike will have questions ready to go, but will also have the mental flexibility to take the discussion in unexpected directions. Importantly, journalists and students will also have good ideas of their own, and they won't be afraid to push back when they don't understand, or don't agree with, yours.

That raises an important point though: journalists *aren't* graduate students. Some of us have been in the past and most science journalists I know are plenty smart enough for it, but that's not what I mean. Your students walk into class owing you a measure of respect and courtesy—deference even. But it's an important part of any journalist's work to actively avoid deference of any kind. The power and value of journalism, after all, comes from the independence of journalists. (When political reporters don't ask tough questions, we get phantom weapons of mass destruction; when science reporters don't push back, we get cold fusion.)

That doesn't—or at least shouldn't—mean disrespect. Journalists, ultimately, work for the people who read what we write and watch and listen to what we broadcast. And

BOX 9.1
Continued

they expect us not just to understand, explain, and contextualize your work, but also to challenge your assumptions, test your conclusions, and get past your talking points.

I wondered though, what makes a good interview from the scientist's point of view? "It really helps when you can tell you're talking to a person, not just a pencil," was the answer one source-turned-friend gave me.

Maybe in the end, it really is that simple. What makes the very best interviews "magic"? You don't need to be friends, nor even always friendly. But humanity, and a dose of mutual respect, can turn a simple transfer of information into an enriching and deeply enjoyable experience. Which, come to think of it, is really nothing like a blind date or job interview at all.

—Thomas Hayden, author of *Sex and War*, is an independent journalist who has worked at *Newsweek*, and *U.S. News & World Report*, and teaches science journalism at Stanford University.

Offer Constructive Feedback

Journalists are hungry for feedback on their stories. If you honestly think a journalist did a good job, write a quick e-mail, say so, and be specific about what you liked. This will help open the door for future positive interactions. But your praise must be sincere.

If the journalist made an error, you should also politely let him or her know. Balance the criticism by noting the positive aspects of the story as well; in short, approach it as you would giving feedback to a colleague or student. Journalists find it extremely frustrating when scientists complain to others about an error but did nothing to call it to their attention. This perpetuates the error because it stands uncorrected. If the error is factual and in the mainstream media rather than the blogosphere, the venue will likely print a correction. This is especially important now that stories are archived online.

Bear in mind that in the editing process, editors often strip out nuance and context in the interests of brevity. Journalists struggle with this daily. Sometimes they win, but often they don't. Also, headlines are not written by the reporter, but rather by a copy editor. So if you don't like the headline, chances are the journalist didn't either.

Thorny Issues

Scientists who are interviewed by journalists frequently struggle with a common set of issues. In this section, journalists as well as scientists offer their perspectives.

I'm Not Quoted in the Story!

It can be frustrating to take time out of a hectic day and not see your words quoted directly in an article or make the broadcast. But that is a shortsighted view. The best reporters often over-research their articles and simply cannot squeeze in all of the voices in the compact time or space allotted for any particular article. That doesn't mean your time was wasted, your views were dismissed, or you didn't make a difference in the way the story came out.

"What scientists don't always realize is how their words, even if not quoted verbatim, influence how I write an article and help me get it right," says Ken Weiss, an environmental reporter at the *Los Angeles Times*:

> I feel forever indebted to these unmentioned researchers who help me avoid mistakes, synthesize complicated ideas, or put things in proper perspective. When colleagues ask me for an expert on a topic, I recommend one of these unsung heroes for quoting. They are usually the first experts I go back to when I do the next article on a related topic.

Pam Matson, dean of the school of Earth sciences at Stanford, says:

> I actually like to talk to journalists who don't necessarily want to quote me. I like to help them get the story right. I get quite a few calls just to talk through where the science is going, recognizing that anything I say could be used, but it's more about the background content.

Justina Ray, the executive director of the Wildlife Conservation Society, Canada, reflects, "I know that what is important is the story in its entirety and not just my specific quote. If the general story keeps with the message, I want to convey then I feel satisfaction."

Reporters appreciate scientists who can see the big picture. So take the

long-term view, think about the greater good, and focus on building relationships. It will pay off.

On or Off the Record?

Nothing gets both scientists and journalists more agitated than the topic of
what is, and is not, "on the record." There is no simple answer. Every journalist and every organization has a different policy. It would be ideal to know the
journalist's rules up front. Unless you have a prior relationship with that person, however, that's not always possible. So I tell scientists, "If you don't want
to see it in print, don't say it."

   This makes some journalists annoyed with me. Ken Weiss of the *Los Angeles Times* thinks this advice needlessly constrains scientists and does journalists a disservice by making scientists overly cautious. But the problem is, every
journalist has a different point of view.

   Here are a few examples, taken from some of our communications
workshops:

> Cornelia Dean, science reporter, *New York Times*:
>
> > When people say "off the record" I stop them. If I cannot use it, I don't
> > want to hear it. If I cannot say where information came from, it is use-
> > less to me. What am I supposed to tell readers? "It came to me in a
> > dream"? I want my story to have credibility with my readers, and it
> > won't if I can't tell them where the information is coming from. You
> > should be thinking way ahead of time about what you want to get out
> > to the public, and then you won't have to worry what is on and off the
> > record.
>
> Jeff Burnside, television reporter, NBC Miami:
>
> > There is often a misunderstanding that you think you are on the record
> > only when the camera is on. That is simply not true—it's always on the
> > record. Life is on the record, if you say something you say it. I cannot
> > use information on background unless I confirm with other sources.
>
> Christopher Joyce, science reporter, NPR:
>
> > Define your terms with the reporter and assume you are on the record.
> > On the record means I can use it on the air with your name attached,

off the record means I can't. There are shades all the way in between. I don't often have to use these terms with scientists. Sometimes I tell people who work in government, "I'll quote you as a government official."

Thomas Hayden, freelance reporter:

> When we have a relationship, it's one of source and reporter. We might be friendly, but we're not friends. My responsibility is to my readers, not you. My goal is to convey the facts accurately. Most times this is reporting the facts as they are told to me, but sometimes there is more to it. A source working in Brazil, for example, said that if they were quoted directly they would no longer be allowed into the country. So if more harm than good will come, I try to accommodate.

Okay, you get the picture. If you want to negotiate, do it up front—there is no retroactive bargaining. But still, it's risky.

In 2005, I led a communications workshop for scientists at the Smithsonian Tropical Research Institute (STRI) in Panama. At the training, Ira Rubinoff, then the director, asked me and the journalists I had invited to explain "on and off the record," which we did. Cory Dean told him, in no uncertain terms, that as far as she was concerned there was no such thing as off the record. The scientists and journalists had an animated debate on the topic, but the take away message was, "nothing is off the record."

Later, Cory decided to do a "Scientist at Work" profile on Jeremy Jackson, a marine ecologist and paleontologist. As a standard step she interviewed a number of scientists, including Ira Rubinoff, for their take on Jeremy. Her story, a portrait of a brilliant and sometimes difficult scientist, included quotes from Ira Rubinoff as well as Biff Bermingham (now director of STRI) who had also attended the workshop (Dean 2005): "Dr. Jackson is a highly proficient scientist who tends to be intolerant of colleagues who don't see things the way he does," said Dr. Ira Rubinoff, who brought Dr. Jackson and Dr. Knowlton to STRI. "His big ego can be very hard to take," said Dr. Eldredge Bermingham, an evolutionary biologist who is deputy director of STRI, "but I admire his research, I admire his passion, and I admire his willingness to speak out."

Jackson was outraged, but not by the comments. "Ira was interviewed by the *New York Times* and he squandered that opportunity on dissing me," Jack-

Cory Dean and Ken Weiss interview Jeremy Jackson. © 2005 Jeff Burnside

son says. Don't say it if you don't want to see it. You can read the full story on-line on the website.

Reviewing the Story

Another charged topic is whether scientists who have been interviewed can see the story, or at least their quotations, prior to publication. As with the "on the record" question, there is no uniform guideline, but for most journalists the answer is a nonnegotiable "no." In fact, for some organizations like the *Los Angeles Times* and the *New York Times*, it's a firing offense for a journalist to show the story to a source before publication. The editors are the only legitimate reviewers before the story is published. Because science is founded on the peer-review process, it's hard for most scientists to accept this. But it is deeply ingrained in the culture of journalism and journalists see it as critical to the integrity of their profession. "Everyone wants to change their quotes. It's human nature," says Juliet Eilperin of the *Washington Post*. "I've been interviewed too, I know what it's like."

BOX 9.2
Why You Can't Read My Article ahead of Time

Juliet Eilperin

From time to time, a delightful interview I've conducted with a perfectly lovely scientist ends in disaster. It always begins with the same question—"Can I see your article before it comes out?"—followed by my equally predictable response—"No."

I can offer up plenty of practical reasons for this policy. Newspapers, especially now that we've entered the Internet Age, operate at a frenetic pace. Most of the time I write articles in a single day, though I occasionally turn around a piece in a matter of hours, and other times I can devote a week or two to assembling a story. No matter how long it takes me to produce a story, however, the editing process inevitably takes place at the end of the day, in a rush. I ship the story over to my editor, who reads through it and sends it back to me with questions and suggested changes. After I go through the article a second time, I send it back to my editor, who in turn ships it over to a copy editor. The copy editor typically reads my piece in the evening, and may ask me a question or two before it's printed. When would I have a source take a peek at my story? At 9 p.m., when the copy desk is poised to send it to our printing plant? Earlier in the evening, when my editor hasn't had a chance to look at it? There's no right time to share a story until it appears online at the end of the night.

Beyond the practical considerations, however, there's actually a philosophical reason why I balk every time a source asked me to share my story in advance: without journalistic independence, my work has no value.

People try to influence reporters all the time: politicians, lobbyists, and, yes, even scientists. It's understandable that individuals with a stake in a specific outcome, whether it's an economic or policy one, would want to ensure that the press paints them in the most flattering light. But it's my job to make sure that once I collect information from a range of sources, I assemble a story free of outside influence. If I did actually show a scientist my work, it's impossible to envision a situation where he or she wouldn't want to change things a little bit—or a lot. People always try to change their quotes when you read them back to them, not necessarily because they want to change the meaning, but because they want to sound more polished, or more bland, or what have you. That sucks the life out of a piece. And in some instances, researchers want to do much more, because they don't like the overall structure or message of an article.

In the end, the only people who get to scrutinize my work before it's published sit inside my newsroom. I'd say that's why they get paid the big bucks, but that's a lie. It's just because they're members of the Fourth Estate, who at the end of the day serve the public, and no one else.

—Juliet Eilperin is an environmental reporter at the *Washington Post*.

I Can't Believe They Quoted That Guy

Journalists are trained to look for and present a range of perspectives. Scientists often feel wronged when those opposing quotes are taken from someone who would not be recognized as a reputable expert by the scientific community.

But how are journalists to know who qualifies as an expert on any given subject, especially if they haven't covered it before? Head this problem off at the pass during your interview. By offering the name of a known critic and perhaps even explaining what their criticism is, you demonstrate a fair and balanced perspective that may encourage the reporter to contact you in the future. It also gives you an opportunity to counter the criticisms gently.

BOX 9.3
The Question of Balance

Kenneth R. Weiss

Why do journalists seek out opinions? Why do they insist on representing opposing views even when it results in promoting professional contrarians or discredited cranks?

The answers vary. Sometimes there's an agenda afoot. But usually it can be chalked up to journalists' training to present all sides of the story, their deep suspicion born from experience, or their skimpy understanding of most scientific fields. This lack of in-depth knowledge makes it difficult, if not impossible, for journalists to distinguish mainstream thinkers from the outliers.

Journalists are generalists by nature. Most newspaper, radio, and television reporters cut their teeth by covering crime, politics, and disputes of all kinds. In covering courtroom action, or a debate at City Hall or Congress, journalists routinely witness lawyers and politicians ignoring inconvenient truths while presenting their arguments. Reporters learn to rush to "the other side" to gather opposing views. Failure to do so would be unfair and unprofessional, and subject the journalist to withering criticism from an editor. If lopsided coverage gets broadcast or published in the newspaper, the journalist often hears about it afterward from public complaints, letters to the editor, or demands for corrections.

Journalists also know from history that it's important to pay attention to lone voices. It was, after all, scientific heretics who challenged the status quo with new ideas that the Earth was round, that refrigerants opened a hole in the Earth's protective ozone layer, that stomach ulcers can be caused by bacteria.

BOX 9.3
Continued

When thrown into technical or scientific arenas, self-critical journalists distrust their ability to distinguish genius from the ridiculous. So they fall back on the crutch learned in the courtroom, offer "both sides" and kick it to members of the public to decide for themselves. It's a safe approach, albeit not necessarily the best one to help advance public understanding of science.

Journalists with some expertise can find themselves second-guessed by editors who insist that an article represent opposing views, even if it provides an unwarranted platform to silly or disproved ideas. To prepare for such demands, science writers and other specialists seek out alternate views but then attempt to put them in proper context. That usually means giving the contrarians far less prominence in the article. Instead of the standard journalistic approach of invoking critics of any merit in the first few paragraphs, the more questionable criticisms are characterized as less than mainstream and weaved into the article closer to the end than the beginning.

So what can a fair-minded scientist do to navigate these journalistic practices? First, understand that it's a common journalistic standard to represent various views. Think of it as a sort of layman's peer review. Second, accept that most journalists do not have the time or expertise to judge for themselves what's right and what's absurd. Help the inquiring journalists put your ideas into proper context. Briefly explain the consensus of scientific understanding, where your idea fits into this, and disclose the principal areas of honest debate. Offer a name or two of respected scientific colleagues who push an alternative theory or who might offer a thoughtful critique of yours. That will enhance your credibility in the journalists' eyes and distance you from their experience with slick attorneys and politicians. It will also help the journalist weigh any counterargument appropriately, which is likely to be reflected in whatever gets published or broadcast.

—Kenneth R. Weiss is an environmental reporter for the *Los Angeles Times*.

A Summary of Do's and Don'ts

> If you work with media a lot—sometimes it will be good, sometimes it won't. Get over it.
> —Stephen Schneider

Here is a short list of do's and don'ts to reference when you have precious few moments to gather yourself before an interview.

Do

Be Accessible and Responsive

If a journalist calls or e-mails you, contact him or her as soon as you can—in minutes rather than hours. Journalists are always looking for scientists they can trust and can ask for insights on the issues and the players. By being responsive and helpful you increase the odds of becoming one of these.

Know Your Own Headline

Try to sum up your message in a few words or a sentence. Then you can provide more details and back it up.

Engage and Interact

Know what points you want to make. Returning to the tennis analogy, think about how you want to place your shots. Pay attention not only to what you are saying at present but how to bridge to where you want to go next. (Have your bridging statements at the ready.) Ask some questions of your own.

Choose Your Words Carefully

Use warmer words. Instead of saying "amphibians," say "frogs." Instead of "vocalizations," talk about "frog songs." Be emotive and use vivid language. This isn't about dumbing it down. This simply makes it more interesting

Be Ready to Answer: "Is There Anything Else You Would Like to Say?"

This is an invitation to restate your headline. You should have a powerful statement ready. If the interview is live, answering "no" makes for a limp ending. Alternatively, as it is wrapping you can say, "One last thing I'd like to say," and succinctly summarize your take-home message.

Don't

Say "No Comment"

Although you frequently see this in the movies, in real life it's rude and can create the impression that you have something to hide. If you can't offer insight on a topic, explain why honestly—"this is outside my area of expertise," or "we can't say anything until the embargo lifts." Tell the journalist what you *can*, and if you know another person who can help, say so. You want to be helpful, not obstructive.

Sweat the Small Stuff

Try not to obsess about what your peers will think of the story or your quotes. They aren't your primary audience. Look at the story as a whole and don't become upset if there is one little imperfection. Think about how the story might inform real-world action.

The Bottom Line

You have spent years of your life studying your subject; so let your passion and personality show. If you enjoy the interview, chances are the journalist will, too. Remember, a good interview is a good conversation.

Steve Schneider sums it up: "Know thy audience; know thyself; know thy stuff." If you follow this advice, you'll ace that interview.

Chapter 10

FINE-TUNE FOR RADIO
AND TELEVISION

A challenge for journalists and scientists dealing with complex science issues: Say something brilliant simply.

—Gianna Savoie

With print media, readers have a lot of control. They can go back and reread passages, dog-ear pages, or underline favorite quotes. Radio and television, on the other hand, come at the listener or viewer quickly—and once the story has moved on, your choice is to either catch up or tune out.

Broadcast journalists call radio and television "one-pass" media because the audience only gets to see or hear something once. The move to the web has blurred the lines somewhat, with the wide availability of podcasts and video clips that can be reviewed at will. Still, however, the vast majority of the broadcast audience will get lost if something is unclear or if their attention wanders.

For this reason, radio and television reporters' primary objective is to grab your attention and keep it until the end. As with print journalists, there is a wide variety of personal style to account for—some of which has to do with their station or network. No two interviews will be the same, and broadcast reporters all employ different forms of story structure to keep you glued to your seat. As you will see, these things have some important implications for you, if you want to excel in your interactions with radio and broadcast media.

In this chapter, radio and television journalists share their insights to help you shine.

Radio

> In radio, your imagination fills in the details.
>
> —Christopher Joyce

In the United States, NPR just doesn't have much competition when it comes to science radio so let's focus there. NPR currently has about fourteen people on staff at its science desk in Washington, DC. One of them is veteran science correspondent Chris Joyce, who started his career working as a print journalist for *New Scientist* and then migrated to radio. A master storyteller, he enjoys helping scientists learn how to tell *their* stories for radio. You might remember his stories from the *Climate Connections* series, or his coverage of the alleged rediscovery of the ivory-billed woodpecker. In Box 10.1 is what he has to say.

BOX 10.1
Radio Is a Visual Medium

Christopher Joyce

I really need you to listen to me for the next five minutes. Go ahead and pour yourself some cereal, let the dog out, make a sandwich for lunch today. But turn up the volume and *listen* to me while you're doing that, okay!?

I want to tell you a story about how carbon dioxide in the atmosphere is making the oceans acidic. I could tell you that when CO_2 dissolves, it reacts with water to form a balance of ionic and nonionic chemical species including carbonic acid, bicarbonate, and carbonate, and even give you the ratio of these species as well as lots of numbers about how much the ocean's pH has changed since 1990.

But I won't. Because you can't keep all that in your head, especially now that your five-year-old is tugging at your pajama pants demanding breakfast, and because you can't go back and "re-listen" if you miss something that's complicated. Sure, you can do that with a newspaper or magazine. But not with radio. Hear it once, and then it's gone.

Instead, what I'm going to do is get Ove, a passionate and articulate marine biologist, to tell you in person what's happening to the oceans. I recorded him while he was swimming above the Great Barrier Reef so that you could hear him breast stroking and wading across the coral flat, pointing out reef sharks and sea slugs and bleached coral heads. I did that so you could get the feel of being there. So you could hear the urgency in his voice. And mostly, so that you and I, in collaboration, could make a little

---

BOX 10.1

Continued

movie in your head. That's my job in radio. Radio is a visual medium. For five minutes, I hold your attention by creating a waking dream, a mental movie, using sound and my voice and the voices of people who count. People who know more than I or even you. I use short sentences. I describe what I see: the outrigger than hovers nearby on a film of pale blue water, the coconut trees along the shore. You can hear Ove breathing through his snorkel, and the gulls overhead, and the clackety-clack of wind through palm leaves.

I have you now. Sure, I'll drop a few numbers and abstract facts in there, well distributed throughout the five minutes so that you don't have to stop and memorize. That'd break the forward momentum of the movie. Ove will help by explaining things using analogy and metaphor. Reefs are "aquatic rainforests" and their future is "a train wreck we can see coming." A reef is a "spinning plate on a stick," delicately in balance and easily knocked to the ground. If you don't know what stag or brain coral look like, you certainly know what a rainforest and a train wreck and a spinning plate look like.

We're on a roll now. You're eating your cereal, but you're listening closely. I add the voices of other experts; maybe we take a trip to a lab where scientists are growing coral in tanks full of bubbling seawater. As they reach down into the tanks to retrieve a piece of coral for me, they explain that some species are resistant to acidic water, and provide hope for coral's survival in a more acidic ocean.

Somewhere in there, I may hit you with a complicated *idea*, but I won't do it using complicated language.

We end up back with Ove explaining why all this is important. Unlike a newspaper reporter, I've saved some of the best for last, to keep you listening and wondering what will come next.

And then the five minutes is over. You've learned a few things about coral reefs, global warming, and ocean acidification. You actually heard the world's experts tell you in their own voices why this is important. You'll tell your spouse about all this tonight over dinner.

Isn't radio great?

Okay, it's time to let the dog in.

Christopher Joyce is a correspondent on the science desk at NPR. You can listen to his story featuring Ove (Joyce 2007) online.

Chris Joyce's Tips for Radio Interviews

Whatever happens to radio when the world wide web engulfs us all completely, there will always be aural storytelling—it's been around since that first
campfire where the guy who killed a saber-toothed cat with a spear told all
his buddies how he did it. The box the sound comes out of may look different, but stories with sound are here to stay.

To get science across in this medium, scientists can and should do the
following:

- *Use simple language.* There is no such thing as an idea or phenomenon
  too complicated to describe to the layperson—only language that's too
  complicated. Radio is ephemeral and listeners can't pause to figure out
  what you meant while the rest of the story continues on without them,
  or they'll lose the thread.
- *Paint visual pictures.* Use words that evoke the senses, and be specific.
  Don't just think about shape and color, but also texture, pattern, and
  scale. Ben Halpern at the University of California Santa Barbara describes what a kelp forest looks like underwater: "Imagine being able to
  fly through a redwood forest. Kelp grow up to 30 to 40 feet tall, and
  their blades—or leaves—form a canopy that the sun shines through
  creating a cathedral effect. In between the 'trunks' of the kelp swim
  hundreds of colorful fish and invertebrates that come in oranges, blues,
  reds. . . ." Maybe it's not something you'd say in casual conversation,
  but for drawing a listener into that landscape, it's perfect. (Author's
  note: In fact, Halpern carefully scripted these remarks as part of his
  preparation for the interview and using the message box—but he made
  it sound like it was just rolling off his tongue. He did the interview by
  phone, so no one was any the wiser.)
- *Use metaphor and analogy.* They can serve either to describe a complicated process or simply as a visual and memorable way of describing an
  object or event. You may never stumble across a perfect analogy, but if
  you can get the main idea across to the layperson, that's all you need.
  "Antarctic ice shelves crumble off the continent like the edges of an
  overcooked pie crust" works nicely to create the virtual world that we
  want to engage the listener in.
- *Use personal anecdotes.* Describe scenes, places you went to do your

work, what it was like. Haven't you swapped stories with colleagues like this:"Everything was going just fine until the morning we realized we each had a botfly" or, "One thing we just weren't prepared for was a visit from a saltwater crocodile on the reef where we were doing our transects . . ."? Stories personalize your research. There's nothing wrong in letting listeners know that scientists are human beings, pretty much like everybody else.

- *Speak with energy and emotion.* You might have heard the expression "people can hear the smile in your voice." It's true, and for other emotions as well. Let people hear that you are excited about what you do. They will put down their magazine or spatula and listen when they hear someone speaking with conviction and passion. Science is already viewed as dry and emotionless by too many. Convince them otherwise with your voice; if you sound bored, they'll be bored too. Just like animals signaling their fitness via energetic vocalizations, *you* should be signaling the quality of your ideas with the energy you put into your voice.

Now, you might be thinking, "Hey, wait a minute. Most of this advice is about elaboration and color commentary! I don't want to do that—my job is to stick to the facts, hammer home the main point!" Well yes, if you are in a confrontational talk show, perhaps. For many long-form radio interviews that you would find on NPR or the CBC, however, it just doesn't necessarily hold true. Approaching radio with an adamant focus on "staying on message" will most likely result in a dry, stilted interview—frustrating to the host and the audience alike. Though you might have steered an unerringly factual course, the end result will not hold the audience's attention, and you won't be able to count it as much of a real success. You should know what you want to get across, but it doesn't have to sound didactic.

What I need from you is a sense of place, and great anecdotes. When you talk to a radio journalist, remember that most listeners believe in science but are a little afraid of it, or at least afraid of feeling stupid. The clearer and more interesting you can make your story, the more likely you will be to break down the barriers between the scientist and the layperson—and the easier my job will be.

If you are interested in learning more, NPR has published a book called *Sound Reporting: The NPR Guide to Audio Journalism and Production.*

Nancy Baron, Christopher Joyce, and James Lindholm share a laugh during a training interview. © 2009 Ken Weiss

Television

Television is perhaps the biggest challenge for scientists because it is the most compressed and the most focused on superficials—appearance, charisma, and "pizzazz." With television you can forget about conveying the three or four messages from your message box; pick one, and prepare to work really hard for it.

As Peter Dykstra, former executive producer of CNN's Science, Technology, and Weather Unit, says, "It's my job to take your life's work and put it into a sentence. Then I have an editor whose job is to make that sentence shorter." During the last forty years, the time windows available to tell stories on TV have become progressively shorter. For example, in 1968, the average sound bite for a presidential candidate on television was 42 seconds. In 1988, it was 9.8 seconds. In 2004, 7.7. . . . You see the trend (Hallin 1992; Bucy and Grabe 2007).

Television is complicated by the fact that absolutely everything has to have visual appeal. So if you are thinking about television, bear in mind that if there's no video, there's no story. If you have a particularly photogenic field site, take a video camera along next time—some television and online venues will gladly use your footage if the quality is acceptable.

Jeff Burnside is a special projects reporter for NBC in Miami. He is the quintessential television reporter—tall, tan, and blond with a soothing voice. Print and radio reporters tease him for his glamorous presentation, but don't be misled. Environmental reporting is his passion. In a past life he also trained policymakers to deal with the press. Burnside tends to disarm you with his soft manner and then leans forward, microphone in hand, to ask a zinger question.

BOX 10.2
"The Bottom Line Is . . ."

Jeff Burnside

You see? It worked. That phrase, "the bottom line is . . . ," gets your attention. "The bottom line is this" should be followed by that little quote or sound bite you have rattling around in your head like an index card waiting to be pulled out and used without missing a beat. But, equally important, using this phrase forces you to think in summarizing terms, in brief overviews, or in concise quotable words that are essential in television.

Having prepared thoughts and responses to your work also helps when you get a call like the one I made the other day. I've done it hundreds of times. It goes something like this:

> "Hi. This is Jeff Burnside. I'm a reporter for NBC6. I'm doing a story about [fill in the blank] and I'd like to get together with you briefly to do an interview."
> "Uh huh. How about Tuesday?"
> "Tuesday? As in 'next week?' No. I'm thinking of forty-five minutes from now. In fact, I'm already on my way over."

That's the kind of deadline many reporters face—certainly most TV and daily print, radio, and online reporters.

My day begins at 9:30 a.m. I walk into the newsroom and immediately go into a meeting to discuss my day's assignment. As soon as I can, I leave so I can get started. I need to get my hands on basic information, search our archives for information and video, and decide who I can call to line up an on-camera interview or two. By the time I am able to reach someone, it's 11:00 a.m. I have seven hours to go until news time. In this business, you are on the air regardless of what you have or don't have.

I finally convince someone to appear on camera. But they are a thirty-minute drive away. So while we drive, I continue to research by phone or laptop. We set up the

---

BOX 10.2

Continued

camera and lights and begin the interview. It's 11:45 a.m. Then we spend thirty minutes shooting video to supplement our story. Now mandatory lunch for the cameraman. Union rules.

At 1:15 p.m., we drive toward our second interview. We finish that by 3:00 p.m. Three hours to go and I still have to review the interview, pick the sound bites, write the script, wait for approval, assemble the story itself, and feed it back to the station.

I am writing the story and selecting the sound bites in the news car while the photographer and engineer set up the live microwave or satellite truck for the live report from wherever.

The photographer and I assemble the story in a mix of video and sound and reporter narration. We feed it back to the newsroom's mainframe computer at 5:50 p.m. I have twelve minutes to prepare for my live portion of the story. The bulk of the story, of course, is finished and beaming back to the station.

At 5:57 p.m., I must be in front of the camera for a sound check and to make sure the frantic producer knows I'm ready and technically good to go. Three, two, one. . . .

My story is finished at 6:04 p.m. Tomorrow, I hope there will be no traffic jam.

So, when I respond "Tuesday? As in 'next week?'" with a bit of incredulity, you'll know why.

—Jeff Burnside is a reporter for NBC Miami.

Jeff loves helping scientists "up their game." Here is a summary of what he teaches in our workshops.

Jeff Burnside's Tips for TV Interviews:

- Even though there's a camera staring at you, be yourself—although it helps to kick up the energy and volume a little because it is naturally diminished on-air.
- Use short sentences with conclusive, definitive endings. Leave a pause between ideas so we can edit cleanly.
- Frame your answers with part of my question. Don't just say yes or no.
- When I interview you, look at me. Not the camera. Think of it as having a conversation with me.
- Don't be afraid to tell me how you feel about something, or why you care.

- Interviews are helpful for facts. But, more than likely, I already got them from you. So the on-camera interview is vital for your observations, thoughts, opinions, and emotions: your sound bite.
- If you don't like how you answered, just say so and ask to answer it again. Or, you can ruin the take by waving your hand in front of your face and saying, "I flubbed it, let's do that again."

Many scientists are disdainful of television. But it has a massive audience, which can work to your advantage. Another benefit is that viewers do not know what's next, so unlike the news they get online, or read on paper, they can't act on a predisposition to follow only news in which they are interested. TV is a smorgasbord, and stories by reporters like Burnside, who are serious about getting important environmental news across, are in the mix with everything else. TV reaches people who might never otherwise find these stories. Perhaps TV may only gloss the surface, but consider it an opportunity to catch people's attention, make an impression, and perhaps create an appetite to know more.

As Paul Ehrlich of Stanford University likes to say, "TV is the most ephemeral of media. This holds true for both gaffes and brilliant sound bites." So relax and try to have fun with it.

The Bottom Line

Because radio and television allow the audience to see and/or hear the story, you have to consider much more than your words in an interview with a broadcast reporter. Enthusiasm, energy, and emotion—or a lack of them—will translate directly into the final product. Spend a little extra time preparing for broadcast interviews, and always be on the lookout for vivid ways to tell a more compelling, visually oriented story. Perhaps most important, remember to be yourself—only maybe a little bit more so.

Chapter 11

REACH OUT INSTEAD OF WAITING

If you aren't going to play an active role [getting your message out], then you can't complain when you don't like the coverage.

—Marla Cone

Don't just wait for that call. There are many ways that you can give your science legs—meaning that it walks into the world and has an impact. This chapter and the next explore a range of approaches that will make you a player: from writing op-eds and blogging to attracting media interest in a talk you are giving, pitching your story to specific reporters, or trying to ensure that a paper you are about to publish gets noticed beyond academia.

Some scientists think it's unseemly—it can feel egotistical or embarrassing to draw attention to yourself. It's true, if you start communicating your science, you *will* draw attention to yourself. This is not a bad thing. People may start reaching out to you, journalists may call, policymakers may invite you to testify, critics may write blogs or articles, you may be asked to appear on *Science Friday*—one thing leads to another.

"It's exciting when responses to your work come from beyond the usual suspects—the science community—when your research is deemed more broadly relevant," says Larry Crowder of Duke University:

For me, this affirmation led to increasing boldness—go ahead write the op-ed, testify to Congress, work with journalists to elevate important national and international issues. I've spent most of the last five years explaining how

153

marine spatial planning could revolutionize ocean management in the
United States and now it is the focus of the Interagency Ocean Policy Task
Force. Now that's exciting!

So here are ways to get involved and attract attention to your research.
We'll begin with some minor changes you can make to the things you already do so that they are more likely to attract attention and then progress to
the more difficult ones.

Make the Most of Your Next Conference

Scientific meetings are not just for scientists. They also provide good opportunities to connect with journalists and science bloggers. Nearly every major
scientific meeting provides free registration to journalists and offers press
briefings, technology workrooms, and access to press releases and journal articles. These overtures are all part of the organizers' effort to get more coverage for your work and for their meeting.

Sure, some meetings are more successful in these efforts than others, but
it's a good idea to consider media outreach any time you are giving a talk.
Talk to colleagues who have presented at the meeting before; some of them
might have experience with press outreach and will know whom to contact.
Get in touch with your institution's public information officer to see what
they know about the meeting, and whether they can help you connect with
journalists who will be attending or who live in the region.

If you have some results coming out in a new paper that you think might
be especially newsworthy, ask the meeting organizers if there is an opportunity to present your work at a press briefing, as this is the easiest way to get
noticed. If you have a paper that has been accepted for publication, talk to the
journal editors about possible media opportunities at upcoming meetings—
they may decide to time your publication to maximize its visibility—and
theirs.

For example, *Science* will sometimes schedule select papers so that they
coincide with the American Association for the Advancement of Science
(AAAS) annual meeting. In 2007, Ben Halpern was the lead author of a paper
in press at *Science* on cumulative human impacts on the ocean, which he also
happened to be scheduled to present at AAAS (Halpern et al. 2008). When

---

BOX 11.1

AAAS—A Great Place to Meet Journalists

The American Association for the Advancement of Science (AAAS) is the largest general scientific society in the world, and its meetings regularly draw several thousand scientists from all corners of academia. There are bigger meetings out there, to be sure, but few better places to bump into international journalists from the *Economist*, the Australian Broadcasting Company (ABC), and the British Broadcasting Company (BBC), as well as top-flight U.S. and Canadian journalists.

All told, AAAS draws nearly 1,000 science journalists, bloggers, editors, and public information specialists every year. Although some journalists voice skepticism about the "newsworthiness" of AAAS because talks are submitted nearly a year in advance, they still like to come to network and to sniff out stories. AAAS has an exceptionally well-run press operation, offering a busy schedule of about thirty press briefings over five days. Hundreds of original stories are filed from the meeting every day, many of which get reposted or rebroadcast by other venues, amplifying their reach even further.

For more details on how to get the most out of this meeting, including how to put together a standout proposal and make the most of a press briefing, look online. 🖱

Halpern helped *Science* and AAAS make the connection, the paper was rescheduled to come out during the meeting. He got a press conference, and the next thing he knew he was on *Science Friday* talking to Ira Flatow.

So the bottom line is, if you have a paper coming out that is related to a talk you will be giving at a meeting, be sure to let both the meeting organizers and the journal know. In these situations, everyone benefits from the shared opportunity to promote your work.

Approach a Journalist about a Story

In journalism parlance, presenting a story idea is a called "making a pitch." There is good reason for you to take the time to pitch your research or a story idea to a journalist yourself. For one thing, journalists sometimes ignore public information officers (PIOs), particularly those who constantly approach them with institution-promoting items that aren't really newsworthy. They are more attentive to scientists because they know it's rare for you to reach out, and they like it when you do. Even if you have a stellar

PIO or communications team, sending a pitch yourself gives them a sense of you and shows you are responsive. It is the first step in building a relationship.

Although everyone has personal preferences, e-mail is generally best because it allows you to compose your thoughts succinctly, and it allows the journalist to reply after he or she has met any pressing deadlines. Keep in mind that when you are typing an e-mail or about to pick up the phone to call a journalist, you need to be brief and straightforward. Journalists are deluged with information, so cut to the chase.

The header is important. It needs to be clear it's a personal e-mail sent to them, not spam. Be as specific as possible rather than general and don't shy away from superlatives if you can support them (for example, the first, the only, the largest, and so on). The unexpected works well too as in this real example: "Scientists burning down the Amazon." It grabs your attention and demands further explanation.

The first line or two of your e-mail should make clear what story you want the journalist to write, or what you are inviting the journalist to join you in doing. Then very briefly flesh out the details. Busy journalists screen their e-mail super fast, so if you do not make your point by the second or third sentence, it's game over.

Next, give the journalist a sense of who you are and why you have an interesting perspective on the issue. If you have met them before you can say so, but move on fast. For example, "We spoke, albeit briefly, at the AAAS Marine Mixer. I am following up. . . ."

If it makes sense, mention others who are knowledgeable about this issue, have an outside perspective, and aren't directly involved in what you are doing. This is analogous to suggesting reviewers for a manuscript. In the interest of keeping that first e-mail short, you can say you can provide this information if the journalist is interested.

It never hurts to have what journalists call a kicker—a last shot that shows how urgent and important your story is. Say something like, "This is an opportune time because . . ." and then convince them. All in all, your e-mail should be very brief—no more than a few paragraphs.

A final note: it is always useful to give journalists as much time as possible. There's nothing wrong with giving them four or six weeks advance notice. This enables journalists to make time to do the story.

BOX 11.2

How to Pitch a Story

Here is an e-mail pitch by NCEAS postdoc Jennifer Balch that really worked. Take a look and analyze why it was so effective.

Subject: Scientists Burning down the Amazon. . . .

Dear . . . ,

Given your experience writing about environment and climate change issues, I wanted to see if you were interested in a story about scientists burning the Amazon. . . .

I'm working with the Woods Hole Research Center to gear up for a burn in the southern Amazon at the end of August (26–28). The scale is dramatic. This is the largest experimental burn in the tropics. It's pretty darn exciting, and a bit crazy, to see a bunch of researchers running around burning down a forest. The overarching question is whether these tropical forests will survive the highly unnatural exposure to human-caused fire, or whether people are turning the Amazon into a degraded grassland. We're documenting when these wildfires can happen, and when they do they can burn over 30,000 km^2, equivalent to that deforested each year. And we're producing some of the first estimates of how much carbon comes off these wildfires.

The context is fantastic. Our field site is located right in the middle of the "arc" of deforestation, at the southern edge of the Xingu National Park, on the ranch of the largest soy producer in Brazil, Blairo Maggi. (Maggi is also the governor of the state of Mato Grosso, and was given Greenpeace's golden chainsaw award.) In the nearest agricultural town, at the edge of this frontier, there are agriculturalists, indigenous people trying to maintain their reserve boundaries, American soy farmers buying up property, environmentalists, etc. It's a hotspot of different interests, and internationally everyone is focused on this region to see how to implement a mechanism to compensate for avoided deforestation. Maggi himself said, "If producers were compensated for protecting forests, Brazil would have an army of environmentalists."

But fire is a major threat to the longevity of Amazon forests. Elisabeth Rosenthal's front-page story on Saturday in the *New York Times* on the Kamayurá tribe of the Xingu Park alluded to the severe fires of 2007, and fear of the upcoming burning season. This would be the perfect place/moment to capture a slice of this year's burning season. And bringing it full circle . . . fire is a major contributor to global warming, as our recent *Science* paper emphasizes (see attached). This just came out in April.

Ideal dates to see the burn are this upcoming August 25–28. If you are interested, but

BOX 11.2

Continued

cannot go, we could work out another way to cover it. I have lots of stunning photographs of fire, and can provide more of this specific burn (or video).

Best regards,
Jennifer Balch
Post-doctoral Associate
National Center for Ecological Analysis & Synthesis
735 State Street, Suite 300
Santa Barbara, CA 93101
www.nceas.ucsb.edu/~balch

Why This Pitch Worked

Jennifer customized her note for each journalist she approached and generated a range of stories from the *Economist* to the local NPR station. Her pitch is conversational and a little ironic—she conveyed a sense of herself as someone who would be interesting to talk to and who has a specific take on her science. She used the surprising headline "Scientists Burning the Amazon" to grab attention from the moment her message arrived. In the body, she highlighted the connection to the governor of the region, the timeliness of the piece, and a recent *Science* study she authored, and she piggy-backed on other related events. She offered for the journalist to join her, or to cover it from afar, and she has great visuals. Finally, she linked to her website, which features her work in more detail and in a compelling way.

Results like this don't just happen. While the message may have sounded breezy, it was very strategic. Take a look at the message box she did before crafting her e-mail (see figure on page 159). It makes a difference. If you look at the *Economist* story, you will see how much of this pitch note was included in the actual article (Anon. 2009b).

Build Relationships with Journalists

While it may feel intimidating at first, connecting with reporters is worth the effort. You can do this outside the context of having an immediate story to pitch, because journalists are always cultivating a network of experts they contact for validation, fact-checking, or just to keep their fingers on the pulse of what's going on.

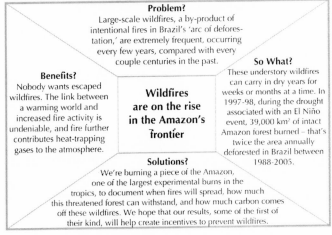

Problem?
Large-scale wildfires, a by-product of intentional fires in Brazil's 'arc of deforestation,' are extremely frequent, occurring every few years, compared with every couple centuries in the past.

So What?
These understory wildfires can carry in dry years for weeks or months at a time. In 1997-98, during the drought associated with an El Niño event, 39,000 km² of intact Amazon forest burned – that's twice the area annually deforested in Brazil between 1988-2005.

Benefits?
Nobody wants escaped wildfires. The link between a warming world and increased fire activity is undeniable, and fire further contributes heat-trapping gases to the atmosphere.

Wildfires are on the rise in the Amazon's frontier

Solutions?
We're burning a piece of the Amazon, one of the largest experimental burns in the tropics, to document when fires will spread, how much this threatened forest can withstand, and how much carbon comes off these wildfires. We hope that our results, some of the first of their kind, will help create incentives to prevent wildfires.

Jennifer Balch's first step was her message box, which led to a story in the *Economist*

Now, saying "build relationships" is easy, but obviously doing it is more challenging. Start with your local media. As a scientist, you are a respected member of the community and they are usually receptive to hearing from you. Pay attention to the news around you. Is there a local NPR reporter you listen to on your way into work each morning? Maybe someone covering controversial environmental stories with insight and depth? Think about the type of issues they cover and whether you can offer something that would be a good fit to those topics. Your first contact doesn't necessarily have to be about a pending publication. You could invite the journalist to tag along on a field trip you are arranging, or to see an interesting seminar or experiment. Whatever you do, make the e-mail succinct and make the offer easy to accept.

BOX 11.3
Invite a Journalist to Join You

Journalists love an excuse to get out of the office. Bringing a reporter into the lab or field with you is one great way to cultivate an ongoing connection and interest in your work. This can work at both the local level and national level, but there are a few things to carefully think through before you make the offer.

- Do make sure that the nature of the story fits the audience of the journalist and the scope of his or her venue.
- Don't overlook your local reporters, especially if you have field sites or student projects taking place around town or campus.
- Do suggest that the reporter might like to meet and hear your ideas about what might be of interest.
- Don't forget about time constraints. While most journalists can manage being away for a day or so, after that it gets much harder to justify the time to their editors. Give them as much lead time as possible to plan for it.
- Do think about breaking a trip into discrete portions. If you are going to be at a remote location or on board a ship for weeks, there needs to be a way for the journalist to peel off after having been there long enough to report a complete story.
- Do think about the visuals—what about your work lends itself to photographs, audio, or video? Is there something strange, exotic, or rare about your site or due to take place during the trip?
- Do plan to make two spaces available if necessary: journalists may need to bring a photographer or videographer along to make the most of the trip.
- Don't worry about the ethics of who pays for the journalists' expenses for the trip. Standards for this vary, but accompanying a scientist into the field is typically not as tricky an ethical question as, say, going to a meeting in Monaco with all expenses paid by a pharmaceutical company. If the journalist seems interested and you can bring them along by all means do so. If they have concerns about costs they will raise them.

If you follow these tips, anticipate the needs of the journalist, and provide them with an intriguing opportunity, you will find you have made a valuable new contact who will continue to be interested in you and your work—even if he or she can't make it this time.

Write Letters to the Editor and Op-Eds

The next logical stage of outreach is to begin writing for the public yourself. There are two main venues for your opinion pieces in print media—letters to the editor and op-eds. Letters to the editor are short (100 to 200 words) and usually address issues or opinions previously featured in the publication. Op-eds are longer (700 to 800 words) and are intended to argue a position or offer a fresh perspective on a timely issue. Both are valuable approaches to adding your voice into a public discussion, but there are some key differences between the two forms.

Letters to the Editor

A print story has a limited shelf life, and only some percentage of the readership reads it. By writing a letter to the editor, you can help sustain a conversation and reach a greater proportion of the outlets' audience. Readers may not necessarily be familiar with the issue at hand, but letters to the editor provide readers with an approachable overview of the original story and reactions to it.

Letters to the editor can also be useful in helping to counteract a negative or contradictory editorial perspective on issues you care about—just remember, you're not writing to argue with the publication, you're writing to reach its audience.

Finally, letters can provide important feedback for journalists. They like positive letters since they mostly hear complaints, but they really do want to know if they have erred. Published or not, letters have an impact—they may change the journalist's mind or educate the journalist. As Cory Dean says, "If I could paint the world differently, whenever a scientist saw something askew he or she would write a letter to the editor. If we have an article that provokes a lot of letters, we all see them. The letters help us."

Of course, outlets can't print all the letters they receive, but the barriers to entry are much lower than an op-ed or a freelance article. Above all, letters must be brief. If you are too wordy, the editors will shorten it to make it fit. If the piece gets edited heavily, you might very well lose your intended message.

Tom Hayden sums it up like this: "Make one point and make it matter. Use wit, emotion, directness, something to stand out. It never hurts to be gracious and acknowledge what you liked too. And should you identify yourself as a scientist? If your point of view is informed by your scientific expertise, yes—but use care with your university affiliation. Be sensitive as to how this will reflect on your institution."

Op-Eds

The term *op-ed* refers to a newspaper page, generally positioned "opposite the editorial page," which features articles expressing a personal viewpoint. Op-eds are a way to reach a broad audience, in your own words, toward your own ends. The best op-eds make a powerful point about a significant issue and prompt readers to reexamine their views.

Op-eds take a considerable investment of time and effort with no guarantee of success. The payoff is that placement in national media like the *New York Times*, *Los Angeles Times*, and *Washington Post* introduces you to decision-makers and the public. A well-written op-ed can help shift public opinion and establish you as an expert. Of course, it is challenging to get published in these national venues but with the web their reach is global. An op-ed in the *Washington Post* may get picked up in Taiwan and Europe.

But your chances of placing an op-ed are best in media where you and your university are known and where you offer an authoritative perspective on an issue of regional concern. Your local or regional newspapers are a good place to start.

Martin Doyle is an associate professor at the University of North Carolina who works on the problem of aging infrastructure, such as dams and bridges. This was not exactly a subject on people's minds until the I-35W bridge collapsed in Minnesota in August 2007. Doyle, who was working behind the scenes with the Army Corps of Engineers and attending public meetings, saw his chance. After learning how to do op-eds at the Leopold training, Doyle wrote and placed several op-eds about infrastructure issues. We talked in January 2009 (see Box 11.4).

Tips for Writing Op-Eds

An op-ed should be a well-argued 700-word idea, not a 1400-word collection of ideas. It must be written in a conversational style for a general audience. The most common mistake scientists make is not presenting their main point at the beginning of the first paragraph. You have to state your premise quickly.

BOX 11.4

Martin Doyle Talks about Op-Eds

You have focused on writing op-eds as a way of moving your issue of aging infrastructure along. What has been your experience?
By publishing a string of op-eds, one a month for three straight months, I have become a player in my state. As a result, I've become a go-to guy for anything relating to environmental restoration and water-related infrastructure like bridges and dams. I get calls from the governor's chief of staff and from state senators, and my name was floated to the assistant secretary of environment. I was amazed at how quickly it vaulted me into that realm of things. And the number of comments I've gotten from the local environmental regulators here in the state on my op-eds has convinced me that this is the scale at which I can make some ripples.

I think my science is unbelievably important. And whether a newspaper covers it gives me some test of that hypothesis. I have been amazed at which of my op-eds are accepted and which are rejected, and that's my empirical database of what the public thinks of my research. In terms of op-eds or pitching something at my local reporter, sometimes there's just no interest.

Have you drawn any conclusions?
Timing is critical. News has to be new, right? It can't be something that would have been interesting a month ago. And I have been amazed at how little interest there was in infrastructure not long ago and how much there is now. But an op-ed has got to be within forty-eight hours of whatever happened.

★★★

When I next talked to Doyle in early March 2009 he had not yet hit a national target, but he had just published an op-ed on the stimulus package in his regional paper. "Not doing bad for the times," he said. You can read Doyle's op-ed online. 🖰

—Martin Doyle is an associate professor at the University of North Carolina and a Leopold fellow.

Op-eds should relate to something that is currently in the public eye, but you do not necessarily have to wait until something happens in the news to place an op-ed. You can, for example, use the publication of a paper as a news hook, but it has to be made relevant.

David Shipley, the editor of the op-ed pages for the *New York Times*, explains, "We look for timeliness, ingenuity, strength of argument, freshness of opinion, clear writing, and newsworthiness. Personal experiences and first-person narrative can be great, particularly when they're in service to a larger idea" (Shipley 2004). Cory Dean adds that you should identify the arguments against your point of view and "demolish them succinctly."

Use thought-provoking and direct language to hook readers' attention. Be concrete and conversational. Rather than discussing "infrastructure," talk about dams and bridges—be specific so that people understand and can visualize exactly what you are talking about.

In addition to making good word choices, you'll also need to use a less formal tone than you are probably used to. Focus on putting things in the active voice and remember this is your chance to voice *your* opinion—don't shy away from saying "I have seen . . . I believe. . . ." Test your work by reading it out loud. Would you actually say that to someone sitting next to you on a bus? Finally, don't assume the audience knows anything about the subject. You must provide whatever context readers need, in as few words as possible.

If you are ready to move beyond your local papers, think about the scale of the issue and determine if your audience is regional or national. It is a good practice to read op-eds where you might want to publish. This will help you gauge whether your ideas offer a fresh perspective in the context of the outlet's coverage. You will also want to pay close attention to the formatting and submission guidelines for each outlet—which can generally be found online.

It is a faux pas to submit to several top-level venues at the same time. Just like with journals, you must wait to hear whether there is interest in your piece before you submit it somewhere else. This can be admittedly frustrating, given the imperative of timeliness. Some journalists recommend you follow up with a call, saying the squeaky wheel gets the grease.

On the other hand, you probably should not squeak at the *New York Times* or the *Washington Post*, since the majority of their op-eds are invited contributions. If you have not heard back after four days, you likely aren't going to, so at that point you can call to confirm that you can move on. "Roughly 1,200 unsolicited submissions come to our office every week via e-mail, fax, and the United States Postal Service," explains Shipley. "Many of these submissions are first-rate—and most get turned down simply because we don't have enough room to publish everything we like."

If the editors decide to run the op-ed, they will contact you to verify that you wrote it. They may edit and fact-check your submission and may shorten it. Don't be offended by that, or by the fact that copy editors will determine

---

BOX 11.5
Top Ten Op-Ed Writing Tips

Juliet Eilperin

1. *Think of your op-ed as a pick-up at a bar, or selling a product door-to-door.* You have a very small window of opportunity to hook your reader, so make your main point in the opening paragraph of your piece.
2. *Simplify, simplify, simplify.* Most readers will know very little about the topic you're writing about. Explain what's at stake in as simple terms as possible (without dumbing it down, of course).
3. *Avoid the passive voice and jargon at all costs.* Passive constructions and insider jargon will turn off a general audience.
4. *Always search for a newsworthy hook.* Is the legislation you care about up for a vote on the House floor? Did it just pass committee? Does it speak to current events—is a whale trapped somewhere, did a massive coral bleaching event just take place? The easiest way to get an op-ed published is to tie it into a recent or upcoming news event. (And no, anniversaries rarely count, unless an editor is desperate for copy.)
5. *If a news event occurs that's tied into your topic, run—don't walk—to your laptop.* An op-ed editor at a major paper will only look at pieces for a day or two after a news event happens. After that, it's too dated.
6. *Paste your op-ed into an e-mail, rather than sending it as an attachment.* That might seem like a minor point, but it makes a big difference with op-ed editors.
7. *Be open to suggestions.* Very often, a newspaper will suggest some changes to op-eds, especially those submitted by first-time contributors. Don't be a prima donna—everyone needs a good editor.
8. *Read a paper's editorial page a little bit before submitting a piece.* In the same way you should be familiar with a newspaper reporter's work before pitching them a story, it's helpful to read an op-ed page to get a sense of what the newspaper tends to publish. That allows you to put your piece in context, either by saying it fits in with the paper's vision or by filling a gap in its coverage.
9. *Don't rely on graphics to make your point.* Some papers, like the *New York Times*, use graphics very effectively on their op-ed pages. But your words should stand on their own—any accompanying graphics should be a bonus.
10. *Above all, be passionate or provocative about what you're writing.* Make sure that you convey the urgency and importance about whatever you're writing about, and above all, make an argument. You are trying to convince people to agree with your point of view.

—Juliet Eilperin is an environmental reporter for the *Washington Post*.

the headline; it is the case for nearly all stories in all newspapers except for some regular columnists. Some media will ask you to sign a contract giving them permission to publish and distribute the op-ed as they wish. Don't sweat the small stuff or slow the process down. Given the competition, you have effectively won the lottery. Whether you are a conservation biologist or chemist, physician or engineer, consider your op-ed as an opportunity to explain why your research questions are important in the first place, why the answers matter, and what you want people to do about them. To summarize, box 11.5 details the advice Juliet Eilperin gives in her tough-love approach to teaching op-ed writing skills to scientists in our workshops.

Writing Online: To Blog or Not to Blog

An obvious frustration of submitting letters to the editor and op-eds has to do with the editorial process. For better or worse, the internet offers the antidote—immediate, uncensored, continuous access to a global audience. When scientists ask whether they should blog, they are sometimes paralyzed: "Will I be wasting my time? Narcissistically navel-gazing? What might people say?" It is probably healthy to consider these questions when your reputation is on the line. On the other hand, the safest route is rarely a useful path for anyone who wants to make a difference. When it comes to blogging, researchers should balance skepticism with a clear-eyed assessment of the power and possibilities.

This section arms you with some advice for finding your voice in the world of social media. Your online presence is inextricably part of your professional presence now. By definition, you are in control. This should feel exciting—blogging is potentially the loudest microphone you've ever stepped up to. The promise of unfettered conversation with citizens, colleagues, and decision makers is heady and scary stuff.

In no time, you can think up a snappy title and find a home for your site. Services like Blogger or WordPress make formatting your site and adding a dozen bells and whistles so simple that you can be lulled into thinking that your biggest concern is "Arial or Verdana"? Congratulations, you have a blog. Now what are you going to do with it?

The process of "finding your voice" involves answering some key questions. What do you want to say in your blog, and what makes you unique? Are you entering a crowded field (climate change) or an empty one (Pacific Island

land snails)? Is your intended audience other scientists, or do you want to primarily play an educational role? Scientists generally find blogs useful for recapping talks or lectures, giving updates from trips to the field, as well as commenting on news, events, and recently published papers. In addition to the education and outreach function this provides, you may find that a blog can

---

BOX 11.6

Simon Donner Talks Blogging

Simon Donner, an assistant professor in the Department of Geography at the University of British Columbia, created his blog Maribo in 2006 (www.simondonner.blogspot .com). He originally just wanted to see what it would take to get people thinking about climate issues and was surprised by some of the benefits his blogging brought him:

> It's a great way of networking. The biggest surprise to me was how many colleagues approached me at meetings—it didn't occur to me that they would read it. And it's a way of working on things you'd talk about at meetings, but that you wouldn't necessarily publish as papers, so it's a great place to test out ideas. On at least one occasion, I wrote something on the blog about policy, got feedback, spruced it up, and sent to a newspaper as an op-ed.

He also comments on the value of blogs in helping scientists address misinterpretations about their work: "If you don't like a press story, or how colleagues respond, you can use your blog to say 'this is our paper, and this is why we did it.'"

But blogging comes with its fair share of headaches. The main concern we hear from scientists is that there simply isn't time. Donner makes a subtle distinction, arguing that blogging doesn't require a large time commitment, but precisely because it is easy and quick, it distracts from the uninterrupted periods of focus that research requires. Furthermore, there is the question of content:

> I think that debates about the science belong in the literature. If I casually trumpet my own work, other people will casually critique it. You have to be careful not to end up in spitting matches. If you really don't think the research is good, write a response to the journal. If you don't do that, you're not contributing to the science community— and you're potentially making things worse for the public too!

Finally, there is the double-edged sword of your audience. Arguments online can spiral out of control, and moderation chores can become a hassle. On the other hand, conflicts draw traffic, so if things are too quiet, it's also problematic. Donner says, "It's a Catch-22—if you don't write often, you are unlikely to have a big readership. If you don't have a big readership, you will question why you are doing it."

—Simon Donner is an assistant professor of geography at the University of British Columbia, and a Leopold fellow.

also help recruit graduate students, build community among far-flung colleagues, and refine the way you think about your current research problem.

Fortunately, dipping your toes into the world of blogging is easy. You might even consider a "roll out" strategy—first test the waters with services like Twitter or FriendFeed (see chapter 5). See how the pace and flow of the virtual world suit you, begin finding interesting communities of people to interact with, and see if your ideas are getting noticed.

It used to be that if you didn't have a website, you were invisible. Now it is even worse, because there are millions of websites, so you're still invisible even if you have one. How do you gain traction and rise above the noise? Use social media and make smart choices. Bottom line: It's not about the technology, it is about the strategy.

- Know what you want to say (what you want other people to get)
- Know what makes your voice unique

BOX 11.7
Writing for the Web

Michael Todd

Flies in amber. Think of them as you write for the web. Each article, each blog post, is expected to be mere ephemeral scribbling, but thanks to Mr. Google and his cousins, they exist in perpetuity, or until the servers crash forever. And just like amber, expect a little polishing after excavation to pretty things up enough if you want these jewels to stay around.

First off, good writing is good writing, regardless of the brooch where it's eventually set. However, there are some cardinal virtues in writing for the web. First among equals: speed. Compared to the glacial pace of newspaper writing, and the geologic pace of academic journals and books, deadlines and news cycles for the web resemble the life of a mayfly.

Speed means what, exactly? Being quick to respond to outside news, for one thing, drafting behind today's headlines to drive home your message. Consider news that touches on your area expertise as an opportunity to enlarge the public debate, and then hop to it.

Speed also requires quick turnaround. This means getting started quickly—often the biggest stumbling block to any writing project—and finishing what you've started. To be frank, half-assed and started almost always trumps perfect and unbegun.

Luckily, two other web virtues reinforce speed: brevity and focus.

BOX 11.7
Continued

In the age of Twitter, which requires posts called tweets of no more than 140 characters, notions of brevity may need some reupholstering; but brief in this instance means from a few hundred words up to maybe 800. (That's about three double-spaced, typewritten pages; it should take you about eight minutes to read it out loud.)

Focus amplifies brevity. Know what you want to say, and home in that point while shedding additional information that isn't absolutely necessary. Being a bellicose sort, I like to use a military analogy: one laser-guided bomb delivered properly is more likely to do the job than a B-52-load scattered in the vicinity.

It generally helps to put the conclusion first, not last, even if that means you're no longer "concluding." Touch on your methodology, don't harp on about it. If inner demons compel you to give more information, consider hyperlinking, to your own work or others'. The same goes for citations—link to them. Like a picture, a URL is worth a thousand words.

Writing tight is not, by the way, dumbing down, nor does it mean losing nuance. Once you've cleared away the clutter of unimportant—or at least not vital—information, the subtleties of the one point you're trying to make can bloom. Such nuance is important too, since the web's barriers to blather are so low that your clarity, authority and context must be in place to raise your message above the chatter.

When stuck for how to be both quick and clear, be explicit. Not exhaustive, but explicit—call a thing by its real name and not by bureaucratic nomenclature, signal that the next point is important or counterintuitive, give a specific example or anecdote. Consider miniatures in an art museum—the detail is exquisite, yet each brushstroke serves to capture a scene in microcosm.

To a lot of prospective writers, the web means attitude. Usually, their definition of attitude is "uninformed bloviating," although they spell it "witty repartee." Attitude is a plus online—if not a cardinal virtue, still not a mortal sin—but it doesn't replace facts, it's a poor substitute for authority, and it's worthless if not in service to your takeaway message.

Keep in mind too that while the advantage of the web is that everyone the world over can see what you write, the downside is that everyone the world over can see what you wrote. So your witty rant can stick to the "you" of the future just as surely as those embarrassing camera-phone snapshots on Facebook do now.

Let's recap. Write short, write quick, and write now. Some attitude is nice, but authority and intellectual honesty will leave you in a better place since everything on the web lives forever, if only in cache view.

—Michael Todd is the online editor for *Miller-McCune*, a print and internet magazine that takes a research-based approach to solving today's problems.

- Find low-cost strategies for building community and audience (for example, make reading/commenting a graduate seminar assignment)
- Make it work for *you*

The Bottom Line

Stepping outside your comfort zone to reach out can have tremendous payoffs. Whether you are building your network of media contacts, writing an opinion piece for a newspaper or blog, arranging a meeting with your local Congressman, or engaged in a media blitz around your latest paper, you'll generate ripples that can lead to surprising and gratifying results.

Chapter 12

PROMOTE A PAPER

The ideal scientist thinks like a poet, works like a clerk, and
writes like a journalist.
—E. O. Wilson

Sometimes, a specific research paper might be important enough that it deserves extra work to get the message out to as many journalists and policymakers as possible. In this chapter, we'll review some important strategies to help you embark on a full-scale outreach effort.

First a cold truth—most science journalists generally only track *Science* and *Nature*, and maybe *Proceedings of the National Academy of Sciences* (*PNAS*). More are beginning to pay attention to *Public Library of Science Biology* (*PLoS Biology*), and nearly all have a handful of journals pertinent to their interests. It might not feel fair or right, but it's the way it is. You can publish, you can even make the cover, your friends and colleagues can toast you, but it doesn't mean the media will cover your paper.

High-profile journals attract media attention in part because they have high-impact factors and strong recognition among the general public. But they don't rest on their laurels either—they put a lot of effort into media outreach. They have dedicated staff who go over the articles published in every issue and highlight what is new and interesting for busy journalists. For example, each week *Science* highlights several papers in a media package called "SciPak," which it distributes to journalists. *Nature*'s press office does something similar. Journalists tend to focus on these "picks of the week."

These sophisticated media packages reach thousands of journalists who are registered to get them, but they are not comprehensive. Even when you publish in a top journal your research can be easily be overlooked. I have seen papers that really should have been news go entirely unnoticed by the press— and that includes some in the most high-profile journals as well those in smaller, discipline-specific journals. A river of news rushes past journalists every week, and unless someone points your paper out to them, it is easy to be lost in the current.

We'll start with a summary of the steps to consider if you want to make the most of your research being published. Then we'll walk through a case study of a paper published in *Science* that probably would not have made news without the authors' intense preparation and outreach efforts.

To be clear: a full-blown outreach effort such as we describe here can quite easily be a month's or several weeks' undertaking and require a small team and hundreds of hours of work. I help scientists do this on rare occasions—when the environmental science is of major interest to society, when

---

BOX 12.1

Is My Paper a Candidate for Outreach?

Outreach efforts are time-consuming. How do you know if the juice will be worth the squeeze? You don't. There are many elements involved including luck and timing, so there are no guarantees. But here are some important considerations. If your paper meets several of these criteria, it may well be worth the effort:

- It is new and surprising, or possibly goes against the grain.
- It can be simply summarized.
- It is clearly relevant to people's lives or interests.
- It is timely and relates to something important happening in the world.
- It has policy implications.
- It is a global result with local examples and stories to tell.
- It has a wide range of authors from different disciplines and countries.
- It will be published in a high-impact journal.

While a high-impact journal is not essential, it helps because they tend to have their own publicity department and your efforts can leverage that.

A paper is unlikely to have all of these elements, and no single one is a dealmaker or -breaker, but the more your paper includes, the more likely it is to get attention.

our scientists deem it worthy (it must be in a peer-reviewed journal), and when the authors are 100 percent committed to investing the time it takes. You can spend money on some of the pieces such as producing video but you can also do it on a shoestring. Sometimes your university has public information officers who can help you write the press release, or tech folks who can do video, podcasts, or the website. It's a lot of work and it's neither feasible for everyone nor appropriate for every paper. But it's not rocket science either. You can do a variation on this full-scale effort and it can make a difference.

In this chapter, we share the steps of our process as a guide so that you can think about whether, if you have the right paper, it's the right move for you. We also offer advice to help identify sources of support from within your institution or professional organization.

Steps for Promoting a Paper

1. Scale Your Expectations and Create a Workplan

A full-blown media outreach effort requires many hours to prepare materials, practice, and engage with journalists, both before the paper is published and hopefully the week after it comes out. It has to have the "right stuff" to make news, but it doesn't end there. You must be committed and willing to do what it takes to prepare. Luck and timing also play a role. If your paper is released during catastrophes or, say, the week between Christmas and New Year's and everyone is on vacation, even the most diligent outreach strategy will only get you so far. Plan to get started as soon as you know that the paper is accepted. Do not make plans to go to Bora Bora or be in the field the week of publication. You have to be totally accessible by phone and e-mail in the week leading up to publication, as well as a few days afterward.

2. Do Your Message Box

Before you do *anything else*—before you even speak to your coauthors or institutional press officers about media work, sit down and figure out your key messages. Ask yourself: What's the story? So what? Why does my research matter? Clarify its relevance in your own mind, and every conversation you have from that point on will proceed much more smoothly. (See chapter 8.)

3. Contact Those Who Can Help

Contact your institution's public information officer (PIO) or communications team. Every institution is different, and the competence and priorities of each press office varies. It is worth feeling PIOs out to see whether and how they can help. Most institutions have experienced staff who can help you draft a press release (guided by your message box), contact journalists on your behalf, and more. It is worthwhile to become acquainted with these individuals so that when you need them they are familiar with you and your work. Good PIOs can be a tremendous help, especially if you give them ample lead time and express appreciation for their work. If they do help you with a press release, make sure to work with them as a team so that the release reflects the messages you want to convey. If you are fortunate enough to have a good press office, ask the staff if they contact journalists directly or just send out the release. Remember that you also can contact journalists directly. (See chapter 11.) In fact, that often brings the best results. Journalists will tend to pay your e-mail more attention than that of an intermediary.

4. Think in Pictures

The old cliché has a lot of truth to it: a picture is worth a thousand words. Good pictures can radically change the visibility of your paper. Journalists will gravitate to stories with good visuals, as it helps their case when they pitch a story to their editor. It can also help you achieve that most coveted of milestones: the cover of a major journal. Many scientists are shocked to learn that it is the art department of most journals, including *Science* and *Nature*, that usually determines which paper to feature on the cover. When it comes to the cover, the best art wins.

As soon as you know your paper is accepted, start looking for photos, graphics, or other art. Look at past issues of the journal for ideas about its tastes. If you think a particular piece has cover potential, you might even try laying it out with the journal's title graphic, because it has to work with the header across the top. We have worked with scientists who have nabbed the covers of *Science* and *Nature* with art that ranged from current and historical photographs to montages and paintings. Felicia Coleman, director of the Florida State University Coastal and Marine Laboratory, thought outside the box for cover art for her study on the impacts recreational fishing. "I took

Diane Peebles, a well-known biological illustrator's beautiful fish paintings and linked them in multiple rows in a PowerPoint slide," says Coleman. "The groupers, snappers, and other reef fish drawings originally appeared on posters commissioned by the Florida Fish and Wildlife Conservation Commission. Diane graciously provided permission, and the rest, as they say, is history."

This cover of *Science* accompanied a paper on the impacts of recreational fishing. (Coleman et al. 2004) Artwork © Diane Peebles. Reprinted with permission from AAAS. All rights reserved.

We once tracked down photographer Gary Braasch in rural Indonesia when looking for artwork to help illustrate Jonathan Patz's study on climate change and disease. Braasch found an internet café, and after numerous attempts on a faulty dial-up connection, was finally able to send a photo that was chosen for cover of *Nature* (Patz et al. 2005).

Braasch offers his advice to scientists who want to help journalists tell their stories with visuals in Box 12.2.

Some journals make photos, maps, and supplementary materials available to journalists. Video clips of interviews with you and your coauthors are very useful, as is supplemental footage to illustrate the story. While you can certainly post these materials to your own website where you feature your paper, or on sites like YouTube, you can (after embargo lifts of course) also put your materials in front of hundreds of journalists by posting them to dedicated sites like EurekaAlert! and AlphaGalileo. (EurekAlert! is a password-protected

---

BOX 12.2
Do You Have Good Pictures?

Gary Braasch

Photographs are sometimes integral to scientists' method. For example, repeat photography helps glaciologists track changes over time, and biologists might use extreme close-ups to identify subject species. Images and video can also fill the important informal need to record and remember what you've done. But more than that, they can make the difference in how and where your story is told.

In today's web-based world, journalists need visuals more than ever before. Print, radio, and television all have online versions that need to be visually compelling. The first question reporters or their editors will likely ask is, "Do you have good pictures?"

These images can not only provide ready visuals for journalists when you publish your next paper, but they can also enrich your relationship with colleagues and expand your connection with the research and university community. When you give slide shows about your work to nonscientists, they will like to see landscape and nature photos that help them understand where scientists work. This kind of image reinforces the idea of science as adventure and discovery. Try to answer the question "so what?" with pictures. Why does your research matter? Pictures make it memorable.

—Gary Braasch is the author of *Earth under Fire*, showing how climate change is affecting the planet. His work has appeared in *National Geographic, Discover, Time, Nature, Scientific American*, and *Natural History*.

website, hosted by AAAS, where journalists can go to find press releases, images, video, and all manner of supplemental resources for reporting on papers of interest. AlphaGalileo is the European equivalent.)

5. Write a Press Release

With the decrease in journalists who specialize in science and the exodus from print media to the web, it's all the more important to write the story you want to see covered (Brumfiel 2009). Sure, the *New York Times* or NPR may not need it because they have the resources and staff to really pursue the news, but that is rare. These days, overloaded reporters are more likely than not to paraphrase a press release; some may even lift parts of it verbatim—especially your quotes. (While many journalists frown on this, it happens all the time.) The better job you do in writing it, the better the overall coverage is likely to be.

Your press release should include the following:

- A headline that grabs attention but doesn't overstate the data.
- Any embargo information displayed boldly at the top, including the time zone of reference.
- A clear statement of what is new and surprising about the work and its relevance to a general audience.
- Pithy quotes—make them memorable and meaningful. A good quotation can go far but a bad one can go viral, so take the time to get them right.
- Quotes from someone not involved in the study who is objective and supportive. Request permission from these individuals before printing their comments and be certain that they have read the original paper. Have their contact information ready.
- Phone numbers and e-mail addresses for lead authors and, if relevant, public relations officers. (You may choose to list all authors especially if you are geographically dispersed as this broadens your appeal.) Include your cell phone numbers if you are willing—it will make it easier to get hold of you at odd hours. Journalists will be grateful, as they would rather go to you directly.

Additional resources for writing press releases are available on our website.

6. Ready, Aim, Fire!

In coordination with your communications office, think broadly about all possible audiences. In addition to the press, public, and policymakers, non-profits and industry groups might be important to target. If the study warrants it, deploy all the tools in your arsenal—prepare a press release, think through the tough questions, e-mail journalists, add a section on the paper to your website, and set up meetings with relevant policymakers. Make the most of the short time when your paper qualifies as news. Unlike op-eds, which you can only submit to one outlet at a time, you can reach out to a multitude of journalists and outlets all at once when you are trying to get the word out about a pending publication. Be strategic given your topic and findings—don't reach out to journalists who don't have a track record of covering your topic. However, be sure to find out and respect the embargo rules of the journal where you are being published. *Science* and *Nature*, for example, both have very strict embargoes and reserve the right to cancel publication if the embargo is violated. Some journals don't have embargoes, but you absolutely need to know from the beginning, and you must comply with any rules, lest you jeopardize your ability to publish there in the future.

A few final reminders:

- Prepare your coauthors. You should agree on the main messages and agree to stick to them. Talk through tough questions you expect to get ahead of time. I recommend you write the questions and answers and distribute them to your coauthors so everyone is on the same page—and if they aren't, at least you'll know about it beforehand. Don't waste your precious few seconds in the spotlight contradicting or tripping over each other.
- Be available! Return those calls and e-mails. Remember that journalists' deadlines are often within hours, not days or weeks.
- Embargo leaks are taken very seriously by journals. Respect the timelines. Obtain assurance from journalists that they will respect the embargo. Journalists who are part of the networks contacted by top tier journals sign agreements to abide by the rules and aren't likely to break them, but other journalists may not know or care. Each journal is different, so know the rules and stick to them
- Set up Google news alerts to track your coverage.

Analysis of a Successful Outreach

Below is one case where the authors of a paper decided that it warranted a full-scale outreach effort, and their investment of time and hard work paid off. This may be way beyond the scope of what you want to take on, and, granted, the scientists had help from COMPASS. But this gives you an inside view to consider the next time you have a paper that you think the world needs to know about.

On September 19, 2008, Christopher Costello and Steven Gaines of the University of California Santa Barbara and John Lynham of the University of Hawaii published a paper in the journal *Science* entitled "Can Catch Shares Prevent Fisheries Collapse?" (Costello, Gaines, and Lynham 2008). The study analyzed the same dataset used by Boris Worm and colleagues in 2006 to predict a widespread collapse of global fisheries by the year 2048. In contrast to that gloomy result, Costello et al. discovered evidence for optimism. By granting fishermen ownership of a fishery's total allowable catch, management systems collectively known as "catch shares" can slow, halt, and even reverse fisheries declines.

This paper was a media hit. Let's deconstruct why.

First of all, it had enough of the right stuff going in. It was a synthetic global analysis rather than an anecdote or a case study. And it was a positive environmental story. However, had we been too literal and just talked about catch shares (what's that?), no one would have cared. It would have been "insider baseball." So instead we framed it more broadly and started with something a lot of people already knew about: that global fisheries are in trouble. We figured that we could catch people's attention by showing them a solution to a known problem. This elevated the relevance of the story far beyond fisheries managers, with an appeal to the general public.

Then we personalized it by focusing on individual locations. This led to substantial international attention. An abstract discussion about a global pattern will only catch some attention; but, if you can make it localized, it will grab more attention in all those places.

It was also timely from a policy standpoint. In several instances, policymakers were struggling to make related decisions, and the authors were able to present their results in this context without overselling what their results meant for possible solutions. Other experts on the topic, potential naysayers, warned against presenting catch shares as a panacea and stressed the importance of customized design prior to implementation in order to fit the social and economic complexities of a particular area. Thanks to this input, the authors addressed these major issues up front, both in the manuscript and in their communications. This helped reduce criticism.

Finally, the study got the right press to have policy impact. Says coauthor Steve Gaines:

> I can't tell you how many people saw it in the *Economist* who didn't see it in *Science*. It was critical in terms of catching the attention of the policy community. That is something I wouldn't have focused on. But if you get something in the *Economist*, it elevates you above the ground-level skirmishing and allows it to get debated at a higher level. Most of the target audience is never going to read the *Science* paper, but if they read the *Economist*, they get the big idea. In many cases, that's the real success here.

BOX 12.3
Anatomy of an Outreach Effort

Christopher Costello

Several months after Costello and colleagues published their paper "Can Catch Shares Prevent Fisheries Collapse?" I spoke with Christopher and asked him to reflect back on the outreach.

What outcomes can you point to as a result of your paper?
This paper received attention almost exclusively because of the outreach and the press. If the paper had been published in a different journal than *Science* and not covered by the big-name media, our colleagues would have patted us on the back and that would have been it—nothing like the response we saw due to our outreach.

Several exciting things have happened over the past year as a result. First, there has been an incredible opportunity to engage in the policy debate. Both Steve Gaines and I served on the Oceans Abundance task force, which made suggestions to President Obama and Congress about ways they can potentially use catch shares in fisheries management reform. We have also worked closely with many fisheries and governments in the United States, Peru, Chile, Mexico, Europe, and elsewhere to help design fisheries management reforms. On the research side, we were approached by several groups to write synthetic articles that

BOX 12.3
Continued

use our *Science* paper as the backbone, but explore other dimensions of fisheries reform and common-pool resource management.

Why did this outreach have impact?

We had a clear and relevant question, got a clean result, tightened up our delivery of that message with your help, got that to the right people, and executed pretty well in the interviews and press conference so that people asked for more follow-up.

Saying the same thing in a slightly different way that translated more clearly to the average person—that was our first take-home message.

Take-home two was the importance of preparation. The preparatory interviews with COMPASS forced me to refine my thinking. What is the bottom line? And anticipating all the things that were likely to come up in interviews.

What have you learned?

Typically, when I publish a paper, as soon as it is accepted and I do the final revisions, that is the last I hear of it. In this case, the outreach literally took more time than writing the paper. Scientists wouldn't like to hear that. But was beneficial to me personally. I spent something like sixty to eighty hours on outreach including preparation on communicating the paper.

One thing I hadn't anticipated is that because of the media attention, other scientists paid attention. And there are two sides to that coin. Some colleagues, who never would have read it, saw it, called me, said, "good job, I enjoyed it." On the dark side, I know that there are a lot of people out there who kind of make a living writing replies to high-profile *Science* papers.

What is your general advice to other scientists?

Practice responding to questions you anticipate. Keep honing the message. I feel like the metaphors we came up with were really important. In our case, there were answers to about thirty common questions. After the first few interviews we didn't hear any new questions. If you can anticipate those questions and be prepared, it's fun and easier to do an interview.

For more details on how they did this as well as links to supporting materials and the resulting coverage, go to our website. 🖱

—Christopher Costello is professor of environmental and resource economics at the Donald Bren School of Environmental Science and Management, University of California Santa Barbara.

The Bottom Line

Intensive outreach on a peer-reviewed publication is a big-time commitment, and your paper has to meet certain criteria. But if you are lucky, strategic, and well prepared, your efforts will be rewarded when you witness the
attention start to snowball—and, in the best cases, you may catalyze meaningful change.

Chapter 13

ENTER THE POLITICAL FRAY

We (academics) are probably the most protected group in the world. Yet there is so much fear. Think of other sectors that do not enjoy such protection, yet they still stick their necks out. We are public servants when it comes down to it. The main reason we have tenure is to protect us from our colleagues. We need to take advantage of that and get out there.

—Rashid Sumaila

You have decided you want to engage in the political arena and prepared what you want to say on a timely topic. Now what? This chapter contains the nitty gritty to help you get started as well as specific instructions on how to make a good impression when you meet policymakers or provide testimony to them.

If you are starting from square one, take advantage of all the resources you didn't know you have. Talk to the government affairs person at your university. The government or public affairs staff of professional societies can also give you guidance and may even be able to help you organize individual meetings.

Scientific and professional societies like the American Geophysical Union (AGU), American Society for Limnology and Oceanography (ASLO), the American Association for the Advancement of Science (AAAS), and the American Society of Civil Engineers (ASCE) can connect you at both the federal and state levels. Many of these groups host Hill Days and arrange visits with congressional offices that will give you hands-on experience with Congress. This is a good way to build a network of congressional connections, which you can draw on to set up your own meetings in the future.

If you want more in-depth training and experience, many organizations offer science policy fellowships. These opportunities are available for scientists at all career stages, from graduate students to tenured faculty, and can fit a variety of schedules, from those with only a few weeks of time to those looking for a sabbatical. Just a few of these programs include:

- AAAS Science & Technology Policy Fellowships
- The Aldo Leopold Leadership Program
- The Geological Society of America's Congressional Science Fellowship
- Harvard University's Science, Technology, and Public Policy Fellowships
- The John A. Knauss Sea Grant Fellowship
- The National Academies' Christine Mirzayan Science & Technology Policy Graduate Fellowship and Jefferson Science Fellowship.

More information and links to the fellowship websites are available on our website.

David Conover and Karen Lips testifying on the renewal of the Endangered Species Act at the Leopold training "mock hearings." © 2005 Aldo Leopold Leadership Program.

Before You Begin

Identify Your Target

As you know by now, your first priority is to identify your audience. It can be a bit daunting to figure out the best people to talk to in the policy world, so when in doubt it's always good to start locally—your representative's or senator's office will make time for you as a constituent.

Despite all of the issues they are dealing with in Washington, members of the U.S. Congress generally spend more than 200 days a year in their home district. This includes weekends, parts of Monday and Friday, and all congressional recesses. Recesses are also called "District Work Periods" and are a good time to meet with policymakers and their staff. The pace of work is slower, and they get to focus more on their constituents. The schedule of recesses is published on the House and Senate websites; they typically occur around the major holidays and the months of August, November, and December. 🖱

That said, don't leave it all on the home front. Take advantage of any trip to Washington, DC, and arrange to meet with your local representatives. You should always try to meet with your own congressman or their staff, even if you are going to the Hill to talk about an issue totally unrelated to the district, or anything they do. It is a courtesy to let them know of the resources (you) within the district, and that you are on the Hill talking to other offices, since that could come back to them in some way. And besides, you may inspire them to get involved.

Consider offices other than just your own representatives. Jessica Hamilton, the natural resources policy adviser for Governor Ted Kulongoski of Oregon, says:

> I would encourage anyone to give me a call. Scientists can also write a brief cover letter, attach research results and request a response from our office. Say, "I'd like to have a meeting or I'd appreciate a phone call." A lot of it may take place on a staff level, but we pass things on to the governor to look at too.

Take stock of who in Congress and elsewhere is leading the way on your issues. This is another opportunity for your government relations and professional society staff to be helpful. Your representative's office might also have advice. With a little time and effort, you can learn to identify issue leaders in Congress (or other policy bodies) on your own. Committee membership and bill sponsorship are solid indicators of interest in a particular issue. Less formal affiliations like caucuses—organized groups of members who share common interests—can also point you in the right direction. Some caucuses, such as the Congressional Wildlife Refuge Caucus and the House National Marine Sanctuaries Caucus, typically have a keen interest in natural science.

To get to the heart of the matter, figure out which committees or subcommittees are most directly involved in the issue you'd like to discuss, then arrange meetings with those committee's staff and members of Congress. Committees allow members of Congress to specialize, which helps them more easily manage the overwhelming breadth of issues they face. Each committee is further broken down into subcommittees with specialized staff.

You can go straight to the websites for the U.S. House of Representatives and the U.S. Senate, but it's often easier to start with a simple web search. This can either direct you straight to the relevant committees or help you find bills that address your issue. If you track those bills through THOMAS (Thomas.gov or Congress.gov), a service of the Library of Congress, you can learn which committees handled them. Again, your professional society or government affairs office may also have resources on their webpage or may have someone on staff that can help answer some questions. Don't be afraid to get in touch with your local member of Congress's office and ask who—or what committee—handles your issue.

BOX 13.1
How Do You Find a Staffer's E-Mail Address?

One handy tip is that even though the House of Representatives and the Senate have different conventions for e-mail, they both have an easy-to-use system of rules to make your life easier. If you know a staffer's name, then you know his or her e-mail. In the House, a staffer named Scott Brown would be scott.brown@mail.house.gov. In the Senate, the address depends on the office. A Joan Smith working for Senator Snowe's personal office would use Joan_Smith@snowe.senate.gov. But if Joan is working in the Commerce Committee, it would be Joan_Smith@commerce.senate.gov.

Every committee has a website that includes its phone number. You can call and ask who on staff handles your issue—just don't forget to ask them to spell the name for you. You can then ask to speak directly to the appropriate staffer, or e-mail them.

Do Your Homework

Before you meet with a policymaker you need to do your homework so you have a sense of what they're working on and where their priorities lie. "If scientists want to get to play, if they want invitations to testify, it's good that they have an understanding of what's on the agenda and how their work might fit in. They should look at our website before they come," Hamilton says.

Here are two basic questions to ask yourself and some thoughts on where to look for the answers:

- **What issues is the *policymaker* interested in?** Start with the policymaker's website to get the big picture view. Most policymakers will list issues and give a sense of their stance. For a committee (or members of a committee), check on their recent hearings, including topics, witnesses, and opening statements, all of which you can find on their website. Look for press releases—they will let you know what an individual or subcommittee is focused on.
- **What issues is the *institution* focused on?** If it's Congress, is it focused on your issue, or is something else taking most of the attention? Tracking the news can give you a feel for this. Don't forget to scan the institution's press releases—for Congress, check the speaker's or majority leader's webpages. You can check both the House and Senate websites for active legislation (the House Clerk is particularly useful: clerk.house.gov). THOMAS highlights particularly active bills and provides access to the official congressional record and other links to help you see what's happening.

Don't neglect outside organizations as a source of information as well. There are advocacy groups who provide "legislative updates" on particular issues, and some nonprofits strive to make things more transparent (for example, opencongress.org).

Schedule a Meeting

Fortunately, you don't need special connections to get started. The process to set up a meeting with your member of Congress or his or her staff is straightforward. Here are the basics:

- Plan well in advance—at least three to four weeks—as schedules often fill up early. Decide if you want to meet in DC or in the district, and look up the recess schedule to plan accordingly.
- Call the member's office. You can find contact information on the House or Senate website, or you can call the Capitol switchboard (202-224-3121) and ask for the office you want. Once you are connected, explain who you are and that you want to request a meeting. Ask for the name of the policy adviser or legislative assistant who works on your issue. Most offices will want you to e-mail or fax a more formal meeting request, but a phone call will get the ball rolling.
- Before you call, be clear about what you want to do in your meeting: introduce yourself as a resource? Share some results you think are policy-relevant? Comment on a piece of legislation? Alert the legislator to a policy or funding need? The more direct you can be, the more attention you are likely to get.
- E-mail or fax your request, making sure to direct it to the scheduler and, if possible, the legislative assistant responsible for your issue. Explain who you are—if you are a constituent, say so. Be clear if the issue is urgent, especially if legislation or another relevant decision is pending, and give a clear picture of what you want to discuss. The more specific you can be the better. "Models of sea-level rise for Narragansett Bay" is much better than "climate change effects."
- Follow up within a few days by e-mail or phone. Schedulers and legislative assistants get so many requests that some slip through the cracks, so if you don't hear back, be persistent. Don't worry about not "making the cut"—most offices have a policy of meeting at least once with any constituent who has a legitimate issue to discuss.
- Once the meeting is scheduled, send some *brief* background information that puts the issue in context. This can be a paragraph or two, or perhaps a compelling graphic or a link to a news article. It will get lost

if you send it too far in advance, so wait until a day or two ahead of the meeting itself. But *keep it short.* You can bring more detailed information to the meeting.

When You Arrive

Make the Right Impression

You don't want to distract decision makers by defying their conventions. It helps to understand their norms. Here are a few things to bear in mind:

- Learn and use appropriate titles. Address Senators as "Senator Smith" or simply "Senator." Members of the U.S. House of Representatives go by "Congressman Smith," or "Mr./Ms./Mrs. Smith." You can find detailed style guides online.
- Don't ignore the dress code. You may have a signature style, but leave the Hawaiian shirts and Tevas at home. Dressing in formal business attire signals to policymakers that you know a bit about their world and value the work they do. It also helps you blend in; the goal is to be remembered for what you say, not what you wear.
- Don't feel offended if those you speak to are constantly checking their iPhones and Blackberries, or if other staff interrupt your meeting. Their schedules can change by the minute, so there are times when they cannot afford to miss an e-mail.
- Assume that you will end up with less time than scheduled. Particularly in Washington, DC, where the schedules are always paced and always in flux, plan on five minutes, hope for fifteen, and dream of thirty. Be prepared for interruptions or to be asked to make your points during a quick walk to the elevator. This means you need to be well rehearsed so that you can make your points quickly. Frontload by giving the most important message first.
- Hand out your business card. Have a one-pager—a bulleted summary of your points with your contact information. (See the next section for more on how to prepare one.) Six weeks down the road, this will help connect you—one of hundreds of people the staffer has met at meetings, cocktail parties, and events—with your message.

- Don't just say, "Let me know if I can help." Be specific—offer to send them a hard-hitting paper or links to a website, and then be sure to follow through. If you suggested another researcher they should speak with, follow up immediately by sending that person's contact information. If it is immediately relevant to your issue, send a formal invitation to show them your lab or field site. Your window of opportunity for follow-up from a meeting is measured in days, not weeks. Finally, take advantage of new developments—particularly those that make the news—and check back in periodically.

Deliver the Take-Home Message with a "One-Pager"

Whenever you meet with policymakers or their staff, you should bring a one-page "leave behind" to help reinforce your points and remind them of the conversation. There are numerous approaches you can take: a well-developed issue that needs attention like the one at right by Joanie Kleypas on ocean acidification.

COMPASS policy director Chad English helped scientists schedule visits on Capitol Hill to talk about the Federal Ocean Acidification Research and Monitoring (FOARAM) Act. The act had been passed, with funding authorized for research, but was still awaiting that next step of actually appropriating the funds. "Chad asked if we had a fact sheet like this," says Joanie Kleypas. "We did not, so he and I pulled this together prior to visiting folks on Capitol Hill, along with packets that included a couple of longer, more official documents on ocean acidification. The busy Washington, DC, folks seemed to appreciate the fact sheet more than the background documents!" Other examples of one-pagers including a personal one (Margot Gerritsen, Stanford) and a summary of breaking science with policy relevance (David Breshears, University of Arizona) can be viewed on our website. 🖱 Although yours may differ in style and substance, it will serve as an effective reminder of who you are and what you have to offer.

How to Prepare a One-Pager

The goal of a one-pager is to help decision makers understand why the issue and science are relevant to them and to remind them of your key points when they look back at it weeks (or months) later. You'll be more effective if you stick to the highlights instead of trying to be comprehensive.

Ocean Acidification – A Major Threat to Marine Life
^ *and Invisible*

Ocean Acidification Fact Sheet

Ocean chemistry is changing

The oceans absorb around 30% of the carbon dioxide added to the atmosphere each year from fossil fuel emissions. This causes several important changes in the chemical properties of seawater, including a decrease in ocean pH known as "acidification".

• This process is well documented and predictable
• This is a direct consequence of rising atmospheric carbon dioxide concentration and is a separate process from climate change.

Ocean acidification has broad impacts

Increased acidity of seawater has a wide range of impacts on marine life. It makes it harder for many marine organisms to build shells, retards the development of larvae, reduces reproductive success, and interferes with some fishes' ability to navigate.

Coral reefs, which support high marine biodiversity and protect coastlines, will slow their growth and even begin to erode in this century.

Marine food webs will be disrupted by these relatively sudden changes. Effects are difficult to predict in detail but are expected to have cascading effects in many fisheries.

Carbon uptake by the ocean: The oceans provide an important climate-regulating service by taking up carbon dioxide from the atmosphere. Over time, the oceans' capacity to absorb carbon dioxide will decrease, but impacts of ocean acidification on biological and geochemical processes makes this even harder to predict.

Solutions

There is no "cure" for ocean acidification other than reduction of carbon dioxide levels in the atmosphere. Until carbon dioxide levels are stabilized, ocean acidification will continue. Ocean acidification is a poorly understood but high-risk consequence of rising atmospheric carbon dioxide levels.

Our ability to deal with ocean acidification relies on how well we can predict its effects. At present we cannot adequately predict how marine ecosystems will respond to ocean acidification. Preservation of our marine fisheries, marine biodiversity, and ecosystem services requires a sound understanding of the multiple impacts of ocean acidification, so that we can help policy makers, fisheries managers, and other marine resource managers deal with these impacts.

How research can help

Ocean acidification is an internationally recognized problem. Major research programs are underway in the European Union and its individual states, Japan, Australia, Korea, China, and New Zealand. U.S. research has been small-scale and piece-meal, but recent passage of the Federal Ocean Acidification and Research and Monitoring (FOARAM) Act illustrates large-scale Congressional commitment to this issue.

The U.S. oceanographic community recognizes ocean acidification as an urgent research priority, and has identified specific research needs. Ocean acidification is relevant to marine chemistry, physics, biology, geology, fisheries, human health, and national security. Research needs will span from large-scale oceanographic expeditions, to space-based analyses, to a wide array of laboratory work. One of the biggest challenges will be coordinating this "web" of interdisciplinary research as efficiently as possible, so that the scientific community can provide timely and sound information to U.S. policy makers and citizens.

Ocean Acidification Subcommittee of the Ocean Carbon and Biogeochemistry Program and Chad English. Used with permission.

• Start with your main point, not with background or context.
• Use formatting, color, and spacing to make your key message stand out on the page so that it can be read at a glance.
• Be concise. It has to fit on one page—preferably on one side.
• Briefly describe the issue in concrete terms. If possible, talk about who will be affected. Use numbers and facts judiciously to illustrate your points—but don't drown them in a sea of statistics. (See chapter 8)
• Give only minimal details of the research—always ask yourself, "How does this relate to the policy issue I'm addressing?"

- Avoid big blocks of text, which make a one-pager hard to read.
- Make sure your contact information is clear and complete. State your name, title, and affiliation. If you aren't at a university, give one sentence about what your organization does and include a link.

Remember, the point of a one-pager is to *start* a conversation and to serve as a reference *after* the conversation. Less is more.

Effective meetings backed up by a solid one-pager can put you on policymakers' maps. They will be more likely to remember you as a useful resource and might call you when they need outside perspectives or expert witnesses for a hearing.

Give Effective Congressional Testimony

Scientists' most visible involvement in policy happens when they testify at a congressional hearing. There are many resources that can help you to prepare effective congressional testimony (for example, see Wells [1996]), so we will stick to the basics here, with a particular focus on science-based testimony.

Hearings allow a committee to get input on a question it is trying to answer. Your testimony will be most useful if you can help committee members to understand their options and answer the questions in front of them.

Before you are invited to testify, a committee staffer will contact you to "feel you out" and determine whether you have what he or she needs. He or she will be listening to see whether you have the right expertise, sound credible, and can talk about your work in a way they can clearly understand. If you are invited to testify, don't be afraid to ask questions. If your role or the questions you are asked to address become at all unclear, contact the staffer for clarification.

Your testimony will take two forms: written and oral. Written testimony is generally five to ten pages and can include figures and references. This is where you can present a complete picture of the science and its relevance to the policy questions at hand. Remember that you're writing for a savvy but nonscientific audience and keep focused on the issues and questions at hand.

You'll need to provide your testimony a few days in advance so that the staffers have time to read it and prepare questions for the members of Congress to ask at the hearing. They might even ask you to help craft these questions. On the day of the hearing, be prepared for specific questions such as

"Should we support this particular bill?" or "What should this committee do about this problem?" It's quite normal to get pointed questions like this, and they can quickly push you out of your comfort zone if you've not thought about them ahead of time.

Oral testimony is meant to provide just the highlights of your written testimony—keep it short and simple. You'll only have five minutes face to face with the committee members, so you need to make it count. This is your opportunity to help them understand the most essential points of your testimony and why they matter. Don't go over your time limit—you'll lose your audience quickly and some committee chairs will simply cut you off. Keep in mind that you are one of four or five witnesses in a row—if each of you gives them a "top ten" list of priorities, then you are expecting them to keep forty or fifty ideas in their heads. Keep your points to a minimum and try to tell a story, using vivid imagery that will stick in their heads.

A hearing is a very formal affair. It will open with a statement from the chair, the ranking member (the highest-ranking member of the minority party), and possibly other members of the committee. Then the witnesses (you) will be invited to testify. After all the witnesses have spoken, the chair will begin "rounds" of questions, where each member gets five minutes to ask and receive answers to questions.

In a two-hour hearing, you might only speak for ten or fifteen minutes in total, but it can feel like an eternity. You can do as you please with the five minutes set aside for your prepared testimony. After that, any time you are asked to speak, it will be in response to questions from a member. When you answer, don't forget that you are using up some of that member's five minutes, so don't ramble. However, your answers give you the opportunity to reinforce your messages. Be respectful of the members' time, and make the most of the opportunity (see chapter 7).

Tips for Testifying

- *Use your resources.* The staff who invited you to testify are busy, but they want you to be effective. Don't hesitate to e-mail or call for clarification on your role, the policy context, who will be attending, and what they'll be concerned about. Ask the advice of colleagues who've done this before.
- *Remember their job.* Make sure you understand what decisions Congress is facing so you can make your testimony relevant to them. The more

effectively you can clarify their choices, the more useful they will find your testimony. They will appreciate direct recommendations, even if they don't take them.

- *Consider your audience.* Make your testimony concrete and tie it to the experiences of the members in attendance and their constituents.
- *Don't read your testimony.* Prepare your written testimony and then boil it down to talking points for yourself. Rehearse until you can make your points clearly while still sounding natural. Stay within the five-minute time limit.
- *Keep it simple.* Don't let the details crowd out your message. Use only as much technical detail as absolutely necessary to make your points. Show you are ready to address the nitty-gritty details, but leave them for the Q&A or follow-up meetings. Statements like "There are some complex interactions that I'm happy to discuss in detail, but the bottom line is . . ." can signal this effectively. Stay away from jargon, and clearly define any terms you can't leave out.
- *Keep visual aids simple.* Make your points using vivid verbal imagery and stories. Be careful about using AV—technical glitches can cost dearly in time and audience focus.
- *Expect them to know your record.* Committee members may ask specific questions about your past writings and any statements you made to the press, so be prepared.
- *Delivery matters.* Be confident, make eye contact, be animated, and tell a good story! Metaphors, stunning statistics, or telling examples will make you memorable.
- *Take advantage of being in the nation's capital.* Since you're going to be in town, make sure you take advantage by arranging related meetings with folks in the agencies and legislature. Without sounding self-important, mention you're coming to town to testify. That will raise their interest in sitting down with you.

The Bottom Line

Don't get discouraged if this seems a little nerve-wracking—it is! Policymakers can be intimidating at first. But if you're prepared, you'll be confident. Overwhelmingly, those scientists who have engaged say it is worth the effort—and exciting. Most policymakers are hungry to hear from you in a way they can understand. The main thing is getting started. Talk to other scientists you know who have already entered the fray. Once you make a few connections, give a few presentations, and show you are able to make your work relevant to policymakers you will be amazed at the ripple effects. People will start coming to you.

PART IV

Becoming an Agent of Change

Chapter 14

AFTER THE SPLASH, THE BACKLASH

Theories have four stages of acceptance:
1. This is worthless nonsense;
2. This is interesting, but perverse;
3. This is true, but quite unimportant;
4. I always said so.

—J. B. S. Haldane

Scientific discoveries are exciting to be sure, but they can also expose researchers to extreme discomfort. Not long after Sherwood Rowland and Mario Molina published their theory that chlorofluorocarbons, or CFCs, could damage Earth's protective ozone layer (Molina and Rowland 1974), industry leaders accused them of spinning "a science fiction tale," dishing out "a load of rubbish," and spewing "utter nonsense."

"The years following the publication of our paper were hectic," says Mario Molina, now a professor at the University of California San Diego, "as we had decided to communicate the CFC-ozone issue not only to other scientists, but also to policymakers and to the news media; we realized this was the only way to ensure that society would take some measures to alleviate the problem."

DuPont, which embraces the slogan "The Miracles of Science," raised questions of scientific uncertainty. The battle became fiercer when the British Antarctic Survey found a hole in the ozone layer. By 1989, the international community came together, recognized the problem, and began phasing out CFCs (Anon. 1989).

Conventional wisdom has shifted since then, of course, and now it seems like so much history. The ozone depletion crisis was solved in record time compared to, say, the ongoing saga surrounding the climate change debate. For Rowland and Molina, it was a wild ride, although it came with a payoff: they were awarded the Nobel Prize for Chemistry in 1995.

"When I first chose to investigate the fate of chlorofluorocarbons in the atmosphere, it was simply out of scientific curiosity," says Molina. "I am heartened and humbled that I was able to do something that not only contributed to our understanding of atmospheric chemistry, but also had a profound impact on the global environment."

The advance of science is far from a smooth and orderly march. If your research survives the normal rigors of peer review, your "new and surprising results" may provoke a strong adverse reaction, especially if they attract media attention. Often, the bigger the splash, the bigger the backlash. The resulting maelstrom can be hard to take if you don't enjoy conflict—which most of us don't.

Backlash results when you have stirred things up. Your research may be threatening to the status quo, to colleagues defending conflicting results, or to a differing school of thought. Interdisciplinary work is especially prone to such conflict as new groups of scientists look into topics that were formerly the exclusive domain of another discipline or interest group. And if you are a "new kid on the block" and identified as the expert on the topic when others have worked on it for years and have a long history of publications, you can evoke personal jealousy. Another form of backlash, which only pretends to be about the data, happens when science comes up against corporate or special interests. This is the politicization of science.

Whether it comes from other scientists or nonscientific groups, and whether the drama plays out publicly in the media or more discreetly in the halls of academia, backlash can shake your sense of fair play. It often ignores scientific codes of conduct, where debates are based on data and methodology. It can turn personal.

The fear that someone will say unkind or untrue things about your work is reason enough for some scientists to avoid such battles. They may decline to offer socially relevant conclusions, or mask them in jargon and caveats so the implications are hidden from all but insiders. "I know so many scientists who shy away from talking about their work—I was one of them—because of fear of the backlash," says Joanie Kleypas, from the National Center of Atmospheric Research (NCAR). "And it's not just the backlash, it's how *other* scientists perceive the backlash when it's from scientists. If they are not experts in the field, then they might perceive any disparaging comments as having some merit."

Yet society needs scientists to provide answers to questions that will help to shape our future, even when people don't like what they hear. In this context, it's important to understand that backlash is part of the process of winning over public opinion and the confidence of decision makers just as it is with other scientists. To deal with backlash you need to anticipate it, be prepared, and try not to take it personally. Recognize it for what it is—the sometimes-painful passage of scientific progress and societal change.

Backlash in the Media

Science reporters thrive on conflict—it's a way to get the public to pay attention to what might otherwise be perceived as an obscure or inaccessible topic. This is why scientific debates aired in the media are often focused more on the conflict and less on the substance. This presents a communications challenge to scientists. While you can say the media is at fault, a contributing factor is that scientists frequently quibble over exceptions rather than acknowledging the larger pattern and what they agree on. Or they may try to ignore attacks and hope they will go away or argue that the critics are not qualified to comment. These approaches do little to win the hearts and minds of the public or help them understand the bigger picture. Scientists need to be proactively communicating on issues of great importance to society on any controversial subject that comes to mind.

Airing debates over scientific details in the media can be destructive because they often convey greater disagreement from the scientific community than may actually exist. This was the case in the earlier stages of climate change debate. While most scientists largely agreed on the big picture, their disagreements on details bewildered the public and policymakers and provided government and industry with an excuse to defer decisions. It also played into the public relations campaigns by special interests designed to instill doubt as a delay tactic.

Andrew Revkin in a July 29, 2008, *New York Times* article (Revkin 2008) says, "Scientists see persistent disputes as the normal stuttering journey toward

"Scientists confirmed today that everything we know about the
structure of the universe is wrongedy-wrong-wrong."

improved understanding over how the world works. But many fear that the
herky-jerky trajectory is distracting the public from the undisputed basics and
blocking change."

When something is reported in the news and there is a lot of subsequent
chatter, the discussion can become distorted, especially in this era of talk radio, cable news, and the blogosphere. So few people read the original source
that it becomes an elaborate game of "telephone" with the message taking on
new interpretations at every pass. Sometimes critics attribute things to scientists that turn out to be unfounded, but they still influence the public discussion. As a consumer of news, it is important for scientists, too, to remember
go to the source and not assume that what is reported is the whole story. It
seldom is.

Kleypas recounts an experience where scientists relied on a media report
and published comments based on it, rather than the original source (Kleypas,
Danabasaglu, and Lough, 2008):

> When we published a paper on something called the "ocean thermostat"
> (the hypothesis that natural feedbacks limit warming in the oceans) we re-

---

BOX 14.1

Advice for Scientists—Not Authors—Asked to Comment in the Media

Controversy is the price of admission in science—but even more so in the public communication of science. However, the rules of the game are different. Science advances by scientists challenging one another, but this happens in the context of peer review or exchanges in the literature. Importantly, uncertainty provides strong motivation to move ahead and resolve controversies.

Outside of academia, controversy is much more public, and the judges and juries often don't understand the history of the issues or the nuances of the science. Opposing views are usually given equal weight. Uncertainty is seen as a reason to tune out or continue with business as usual.

Think about who your audience is—in this case, the public, not scientists.

Talk to your audience—the public, not your peers in the scientific community. Think about your neighbor or a relative. Address why they should care and what matters to them.

Focus on the big picture. Don't get bogged down in the minor details. Often you will disagree with either details of the paper or how it is framed. It's fine to say that but don't lose sight of the opportunity to focus on the real story.

Acknowledge what you do agree on in order to help the public and policymakers understand the broader context and where there is consensus.

Put your energy into constructive ways of disagreeing. Talk is cheap, but this is science and needs to be supported by data, analysis, and peer review.

—This was part of a presentation by Jane Lubchenco, Steve Palumbi, and me to the Pew Marine Fellows. See more on the website.

---

ceived a tremendous backlash from other scientists. I knew the idea of an ocean thermostat was controversial but I work on coral reefs—an ecosystem that has been dropping dead from heat stress—so we examined whether there was any merit to the idea that some regions of the oceans would warm less rapidly. Before even reading the paper, some scientists started to criticize us basing their arguments on a BBC news article (Anon. 2008a). They were livid about the news coverage, because they felt we were perpetuating an old, debunked theory. In reality, our paper and news release both made the point that our research did not show the thermostat-like effect in the future, but some of the news coverage failed to make that point. One scientist even published disparaging comments in a journal article, but his comments were actually related to content on the BBC report and not our paper. He eventually admitted that he did not read the paper itself.

When your science deals with contentious topics and gets into the media, the attacks can come from many different directions. First we'll look at some examples of backlash from within the scientific community, then examples from special interests.

Backlash from Other Scientists

> Science ultimately can only be a never-ending series of approximations towards the truth.
>
> —James Rainey

It's helpful to take a longer view on backlash because although it may be very uncomfortable along the way, things often work out in the end. The following case study examines the history of a big dust-up between marine ecologists and traditional fisheries scientists that eventually led to collaboration.

In 2006, the paper "Impacts of Biodiversity Loss on Ocean Ecosystem Services" by Worm et al.—which became known as "the 2048 paper"—succeeded in two things: getting the world's attention focused on overfishing and raising the ire of scientists who disagreed with the projected collapse of commercial fisheries by 2048.

In 2002, a large meeting of prominent marine biologists working on the Census of Marine Life convened at Scripps Institution of Oceanography to report on the state of ocean biodiversity. Jeremy Jackson, the conference organizer, asked me to invite several journalists who could provide a reality check to the scientists: what would their findings mean to society?

After a couple of days, Natasha Loder from the *Economist* raised her hand. "I've been listening to you lot for two days now," she declared, "and you still have not convinced me that marine biodiversity matters." After a dead silence, the scientists in the room looked at each other and nodded rather than voicing outrage. She was right. They had not made—and could not yet make—a very strong case. This realization led to a four-year study based at the National Centre for Ecological Analysis and Synthesis (NCEAS) led by Boris Worm and Enric Sala, then of Scripps, aimed at answering the question, "Does marine biodiversity matter?" I participated in this scientists' working group, at the end, COMPASS helped the scientists communicate their answers to Natasha Loder's question.

In November 2006, the team of fourteen ecologists and economists published the results of this effort in *Science*. The paper was not primarily about fisheries. It was about species richness and the role it plays in maintaining an ecosystem's productivity and stability. The scenario that projected the collapse of commercial fisheries by 2048 was but a single line in the paper intended to illustrate the ultimate consequences of ongoing biodiversity loss. Still, it answered the important question: so what? Biodiversity is still an esoteric concept for most. The coauthors collectively decided to include the 2048 trajectory (barring one who neglected to review the press release). They felt their data supported it, and it was an example that they knew the world would care about. While the larger message about biodiversity was also well reported, especially by media like the *Economist*, "Fisheries collapse by 2048" splashed across headlines around the world. Coincidently, another paper on the topic of biodiversity, which had considerable overlap, was published only two weeks earlier in *Nature*. It even included one of the same coauthors. Not a single journalist reported on it. It was the Worm et al. study that became news worldwide—because the coauthors worked hard to communicate it and deliberately chose to go with a hot tagline.

The backlash from the scientific community focused largely on the 2048 number, which showed the continuation of the trajectory at the time if nothing were to change. (It was a projection, not a prediction.) Fisheries scientists also took issue with the use of catch data (the reported amount of fish caught globally), which was the most comprehensive data that existed globally. Some felt that the paper ignored healthy fisheries as well as places that are beginning to improve after years of overfishing. The paper's lead author, Boris Worm, became the primary target of critics. Outspoken detractors like Ray Hilborn, a respected fisheries scientist from the University of Washington, called the analysis "mindbogglingly stupid," which was widely reported in the news (Bernton 2006). The schism had much to do with the difference in worldview between fisheries scientists and ecologists.

However, when Worm and Hilborn were interviewed together on a National Public Radio call-in show, they were surprised to find more common ground than they would have thought existed. This eventually led them to bring together an international team of academic and government scientists in fisheries, ecology, and economics for a two-year series of working group meetings, again at NCEAS. The group collected data from every source it could identify that was willing to share it. Layer upon layer, the group collectively analyzed both regional and global fisheries in what it called a Russian doll approach, with one set of data nested inside the next.

In an April 2009 news article in *Science* (Stokstad 2009b), Ray Hilborn stated, "This is the most interesting thing I've been involved in, in a long time." Boris Worm concluded:

> You can already see how things will trickle down and be taken up and processed by the next generation of scientists, who hopefully will not be part of that polarized debate any more. Although it can sometimes be useful to have contrasting views, he says, "there is only one world and we need to work on it together."

When published, the paper also received worldwide media attention (Worm et al. 2009). As coauthor Beth Fulton, a research scientist at CSIRO Marine and Atmospheric Research in Australia, noted:

> It seems to have grabbed way more attention than I was expecting. And many people have been as interested in the social "how did it come about" side as the science side. I'm afraid most of Australia now knows Boris hopes to have a seafood party in 2048 and that ecologists and fisheries scientists can get along.

It is possible to look for constructive ways to move past backlash, to advance the science as Worm and Hilborn succeeded in doing. This is not to say everyone rides happily into the sunset. There are new critiques of the Worm and Hilborn et al. paper and the authors may find themselves on opposite sides of future debates but such is the progress of science.

In the end, you can work with human nature and what audiences will find interesting about a story or you can resist it. The most basic way to make people care is to form an association between something they don't yet care about and something they do care about (Heath and Heath 2007).

"The top guiding principle I have is stay with the facts," says Worm. "Don't let your self say something that you can't back up. If someone presents different facts, seriously consider those and weigh them."

Advice for Dealing with Backlash

While you can't make backlash go away, you can take steps to make sure it does not take you by surprise or hijack your messages. Though it's still hard, you can mentally prepare yourself so that you don't take it so personally and know it's often the price of progress.

Anticipate It and Get Ready

If you know you are publishing a paper that is likely to generate some backlash, be prepared. Anticipate where things could derail. Identify the messages you want to convey and how to express them. Have frank conversations with your coauthors to think through the tough questions, and deliver a consistent message. For the Worm and Hilborn paper, the twenty-one authors worked together to actually write out their collective answers. For example: what was the good news, what was the bad news? This took a lot of back and forth. Writing a press release that all twenty-one agreed to was a monumental undertaking. Worm and Hilborn worked hard to bring the team to consensus. It paid off.

Be forthcoming with information—especially with your critics. Designate a webpage where you can make key documents available—including the paper, a press release, visuals, supporting material, comments from other scientists not involved in the paper, and a compendium of answers to frequently asked questions. If you have the time and resources, you can record video or audio interviews.

It's helpful to talk to other scientists ahead of time—both supporters and potential critics. While it's not always possible to get your critics on board, it's good practice to identify them, learn from them, and address valid criticisms up front. How you frame your messages can be informed by your critics. Many arguments that go public have little to do with data or methods and more to do with how the problem is communicated and whether the results rile someone. So while you can't always stop backlash, you can at least be ready for it.

Consider the Source

Why is this backlash occurring? What underlies the criticisms? Is it honest scientific discourse among scientists who are trying to advance our understanding but have conflicting data or perspectives? Or perhaps it is a perception that what you are saying is an attack on someone else's values or interests. Think about the differences between audiences, their perspectives, and how the debate is aired—through the media or in the scientific literature or at scientific meetings. Consider the best approach for a particular audience but stay consistent with your messages.

Pick Your Battles and Stick with the Data

Dealing with backlash is exhausting. Overcoming backlash requires repeating your messages over and over again rather than backing down. But you don't want to engage in every skirmish. You should ask yourself whether a given exchange is an opportunity for resolution, or will it only serve to further escalate the controversy? If questions recur, address them on your website. Then you can refer people to a well-presented, thorough explanation right next to the data. This approach is both effective and time-saving.

Personal attacks reflect poorly on scientists in general. The best result is when others challenge the research by doing their own analyses to try to disprove your results and then publish in peer-reviewed journals.

When Backlash Takes You by Surprise

Sometimes you may not see the backlash coming until the stories hit the press. When Chris Darimont, a postdoctoral fellow at the University of Santa Cruz, recognized this was happening, he was able to adapt his messages and quickly diffuse the backlash.

Darimont studies ecological and evolutionary processes—natural and human-caused. He and his coauthors published a paper in 2009 showing that humans are causing unprecedented evolutionary changes in harvested populations of wild mammals and fish (Darimont et al. 2009). Because of our tendency to hunt or fish the biggest individuals, we are selecting for slower-growing individuals who mature at smaller sizes.

Darimont did not anticipate that many in the hunting community would perceive the media coverage as an assault on fishing and hunting in general. As soon as this became apparent, he reacted quickly. "I was sure to mention in all subsequent interviews, 'This is not an assault on hunting. Not any more than a study on the perils of pesticides is an attack on agriculture. It's often how we hunt and fish that is the problem,'" Darimont says. He found this approach to be effective. "I could almost see the heads of the reporters bobbing up and down. And I could imagine my uncles—hunters to their very core—agreeing, if somewhat begrudgingly. And almost everyone asked me about solutions. I was glad I had them carefully prepared."

What made the tremendous effort of the outreach particularly worthwhile to Darimont was receiving messages of support from several hunters, fishers, and managers.

Backlash from Special Interests

> You cannot wake a person who pretends to be sleeping.
>
> —Navajo proverb

Most scientists don't go looking for battles. Instead, they find themselves inadvertently embroiled when their science conflicts with special interests. Suddenly, their research and their own credibility come under purposeful attack. This form of backlash is less about science and more about the politics of who stands to gain or lose.

Where financial interests become involved—such as the tobacco industry's denial of scientific evidence showing threats to human health or the petroleum industry's obfuscations of climate change science—backlash becomes more difficult to deal with. You may not win the day against those who don't care whether the science is right. But even if you can't convince "a person who pretends to be sleeping," your clarity can influence others who are trying to make sense of what to believe. Scientists who persevere in these battles become excellent communicators through necessity.

Industry and academic science are obviously not always antagonistic. Industry relies on science too. Many industries support outstanding science and employ excellent scientists. In some cases, though, challenging an industry's status quo and asking questions can open the door to progress.

Even when the immediate backlash is ferocious, there can often be positive outcomes over time. In some instances, industry may fight back in the public arena, but at the same time scientists may become convinced that things need to change and gradually they do. Roz Naylor, an economist at

Stanford University, began examining the sustainability of producing farmed fish when the aquaculture industry was focused largely on issues of production. Today, many industry people have shifted their worldview to include issues of sustainability, but when scientists from outside industry began asking questions such as, "What is the impact of aquaculture on world fish supplies?" industry's first reaction was to lash out at the scientists (Naylor et al. 2000).

BOX 14.2
Dealing with Backlash from the Aquaculture Industry

An Interview with Rosamond Naylor

In June of 2000, Rosamond ("Roz") Naylor, a Stanford economist who studies sustainable agriculture, was the lead author on a review, "Effect of Aquaculture on World Fish Supply" (Naylor et al. 2000). The paper's bottom line was that aquaculture could augment dwindling fish supplies but it's also a problem: whether it is a gain or drain depends on the species farmed. The paper also addressed other environmental impacts of aquaculture and called for sustainable practices that would enhance world fish supplies.

Backlash from the aquaculture industry came fast, and it was furious. Although the paper had ten coauthors, much of the criticism targeted Roz as lead author. I interviewed her eight years after helping her communicate it, for her long view on the outcomes.

How did you deal with the backlash—both professionally and personally?
I was a bit overwhelmed by the intensity of the industry backlash because I didn't anticipate how strong it would be. It coalesced the industry side into first saying, "who are these guys," "why did they do this," and "how can we squash it," then, second, "what are we going to do about it?"

The hostility was generated by the fact that the aquaculture industry hadn't heard of me or this group of coauthors nor did they think it was our business to say anything about aquaculture. The three points of backlash were (1) that they didn't think we had credibility; (2) they thought that we had used the wrong data because they weren't the data that the industry used; and (3) that we had overstated our claim.

The key strategies that I learned worked well were to talk politely and listen to people and to keep on the message. Instead of being defensive, I could keep a perspective that they *didn't* have—namely that when you look *beyond* the human enterprise, to the globe and all the species that are trying to survive with the same resources, it was much easier to reason with them. They did understand that point.

BOX 14.2
Continued

Regarding the data, my point was that if all these data are proprietary—that is, the industry has the data but is unwilling to share them—then we have to use the data that are available and the data that are available are generally FAO (Food and Agricultural Organization) data. If you give us better data that you think are of better quality or more appropriate for the analyses, we will use them. So they eventually backed off that as well.

The overstating the point had to do with the fact that our press release emphasized the finding that aquaculture had a negative impact on ocean ecosystems particularly in terms of using more fish for fish meal and fish oil than it produced in terms of the product for species like salmon. And most of the people who criticized our paper actually never read our paper.

What has been the upshot with the aquaculture community?
I think the upshot is that many people within the industry are trying to figure out how to do things better. They saw that the environmental confrontation was so strong that they weren't going to be able to go forward easily. The *Nature* paper was referenced a lot both at the California level and nationally. It provided credibility for both legislation and for moving things forward on the certification and consumer side, because a lot of press picked it up. They found the article very easy to read and understandable, and I think it got spread around to a lot of people who might never have seen it in *Nature*. It educated the general public as well.

I just didn't anticipate that the industry would take it so seriously. But I think that's the really good thing—that they took it so seriously. And as a result it ultimately moved things ahead.

—Roz Naylor is now working collaboratively with industry leaders and scientists to move toward sustainable practices in aquaculture. For the complete interview, see the website.

The Risks of Speaking Up

When you are a tenured scientist, the risks to your career are more about your reputation. Jeremy Jackson recounts how when he first started reaching out beyond academia, his Scripps colleagues said, "Jeremy's lost it, he's gone applied." For others, job security can be a big concern. Scientists working for government agencies, or young scientists, do not have freedom of speech.

Much is at stake for young scientists who work on controversial topics regardless of how good their science is. This makes mentorship and the support of senior scientists and their academic institutions all the more important. It also means that, at least in some institutions where the support is not there at the senior level, young scientists who are academically top notch may be penalized when they apply for permanent academic positions.

Martin ("Marty") Krkošek studies the role of salmon farms in parasite infestations and declines of wild salmon populations. In 2007, while still a PhD student, he and his colleagues published a series of high-profile publications that ultimately showed that populations of wild salmon were rapidly declining due to the sea lice infestations associated with salmon farms (Krkošek et al. 2006b; Krkošek et al. 2006; Krkošek, Lewis, and Volpe 2005b).

The attacks from industry and some government scientists were intense. Krkošek responded to queries from the press and met with government officials and managers to present his science. But his painstaking response to questions and challenges, media attention, and political cartoons helped his work slowly penetrate the public's awareness. He consolidated everything on his website to provide easier viewing.

Adrian Raeside. Used with permission. © 1998, Henry Dog Productions. Inc.

His work has led to policy change and new standards for managing sea lice linked to salmon farm managements (see chapter 15). In May of 2009, a year into his postdoc, I asked him about the professional fallout. He replied:

> The outreach has impacted my candidacy for academic positions, in both positive and negative ways. I think it is too soon in my career to judge whether or not the outreach has been a net benefit or net detriment. I would say for sure however that it is safer to engage in such activities after securing tenure!

Several universities vied to offer Krkošek a faculty position, but not the University of Victoria in British Columbia where he had hoped to stay. He accepted a position at the University of Otago, in Dunedin, New Zealand.

Boris Worm offers an alternate example that shows speaking up may be valued by your institution. Worm was a postdoc at Dalhousie University in 2004 when he coauthored the controversial paper that came to be known by its assertion that "90 percent of the big fish are gone." Worm is now a tenured professor at Dalhousie, where he has been rewarded for his efforts to make his science matter. The culture of institutions matters, as does the support of other colleagues. This is why it is important for scientists to work collaboratively to encourage and make it safe for colleagues, and particularly young scientists, to speak to the public about their work.

Agenda-Driven Backlash

> You have enemies? Good, that means that you have stood up
> for something, sometime in your life.
> —Winston Churchill.

Benjamin Santer and the International Panel on Climate Change (IPCC)

> The integrity of climate research has taken a very public battering in recent months. Scientists must now emphasize the science, while acknowledging that they are in a street fight.
> —*Nature* editorial, March 11, 2010

Some types of backlash come from those who have political motivations or a financial stake that compels them to refute the science. Often, such attacks have nothing to do with a rational debate about the facts. The goal is to sway the public opinion or muddy the waters so as to postpone or stall logical policy changes that would follow clearer scientific insights, such as the role of anthropogenic greenhouse gasses in changing the climate. Such politicization of science can be brutal.

Scientific debate is like selling a car—here's your baby, you've shined it and tuned it up as well as you can, and then prospective buyers come kick the tires and jiggle the doors to see how it holds up. Maybe they stomp on the brakes a little too roughly during the test-drive . . . but it's all to make sure the car is sound. Agenda-driven industry backlash can be like someone coming after the car with a chainsaw and blowtorch.

Benjamin Santer, an atmospheric scientist at the Lawrence Livermore National Laboratory, has been near the center of the climate wars since the beginning. He says:

> The car analogy is apt. During my early IPCC days, in the summer of 1996, I was receiving some pretty nasty hate mail, and was getting a bit concerned about my physical safety. I used to check underneath my car every morning just to make sure that the guys with chainsaws and blowtorches hadn't been working on my vehicle.

Santer was the convening lead author of chapter 8 in the second IPCC report in 1995: "Detection of Climate Change and Attribution of Causes." "After several years of deliberations involving hundreds of scientists worldwide, we reached the now historic conclusion that the 'balance of evidence suggests a discernible human influence on global climate,'" he says. "Despite its mild wording, the 'discernible human influence' conclusion was significant. This is because it was the first major international scientific assessment to find that our penchant for burning fossils fuels had changed the chemical composition of Earth's atmosphere and hence our climate."

The reaction to the IPCC report astonished Santer. "It was far easier to shoot the messenger than to have a serious debate about the potential economic and social implications of human caused climate change," he says. "It was easier to spread disinformation, to question motives and integrity, to find instant experts who could cast doubt on the science." Santer withstood personal assaults and allegations that chapter 8 was biased, that he had "faked the results," and that the chapter was the product of political tampering. For a time, this unexpected backlash against what he had at first thought was a straightforward scientific assessment took over his life. Subsequent IPCC reports have validated the assessments, and Santer has carried on with his involvement with the IPCC.

On April 23, 2009, Andy Revkin published a front-page story in the *New York Times* that revealed the disingenuousness of Santer's attackers (Revkin 2009). The Global Climate Coalition, a U.S. energy lobby group that had been among the most vocal in challenging the IPCC's "discernable human influence" conclusion, had willfully ignored the advice of its own scientists. A 1995 document leaked to Revkin showed that the coalition's own scientists had written a report supporting the science. "The scientific basis for the Greenhouse Effect and the potential impact of human emissions of greenhouse gases such as CO_2 on climate is well established and cannot be denied," the coalition's own scientists wrote in an internal report.

Yet the coalition advanced a message of uncertainty: "The role of greenhouse gases in climate change is not well understood," they said in a scientific "backgrounder" for lawmakers and journalists issued in the early 1990s, adding that "scientists differed on the issue."

"I was glad to read Andy Revkin's article," says Santer reflecting on his experience:

> It's encouraging and gratifying that the truth finally came out—albeit nearly thirteen years after the events surrounding publication of the IPCC's Second Assessment Report. . . . I think the main lesson I learned is that some things are worth fighting for. The Second Assessment's Report's "discernable human influence" was certainly worth fighting for.

In 2010, climate scientists continue to be in "a street fight" with global-warming deniers who are fueling doubts and dominating the media. As the editors at *Nature* remind scientists:

> The core science supporting anthropogenic global warming has not changed. This needs to be stated again and again in as many contexts as possible. Scientists must not be so naïve as to assume that the data speak for themselves. . . . Scientists can and must continue to inform policy-makers

about the underlying science and the potential consequences of policy decision—while making sure they are not bested in the court of public opinion. (Nature, 2010) 🖰

Barry Noon and the Spotted Owl Controversy

Barry Noon is a professor at Colorado State University's Department of Fishery and Wildlife Biology. In the 1990s, Noon inadvertently found himself in
the center of the northern spotted owl controversy when he was an em-
ployee of the U.S. Forest Service. At the time, he directed a research group
studying the effects of timber harvest, road construction, and other land uses
on fish and wildlife populations in the Pacific Northwest.

His research was distorted when two federal agencies, the Bureau of
Land Management (BLM) and the U.S. Fish and Wildlife Service (FWS)
were pitted against each other. The timber industry lawyers represented the
BLM and took a lead role in the hearings seeking exemption from the Endangered Species Act for critical habitat requirements on federal forest lands
in Oregon. "There were so many half-truths being put forward to support a
given ideology that the public was understandably confused about what the
science was telling us," says Noon. "At that time, I began to view interactions
with the media as great opportunities to communicate the results of the science and to somehow set the record straight."

As the scientist that the timber industry tried to discredit, Noon learned
some hard lessons. "The most important advice I can offer to scientists faced
with explaining science to a hostile audience is to do your homework before
you go on record," Noon says. He continues:

> Address scientific uncertainty by talking about the degree of certainty that
> accompanies our scientific conclusions, not the degree of uncertainty. For
> example, "our research allowed us to state that the species was in decline
> with high certainty." High certainty still admits to some degree of uncer-
> tainty and it's the correct answer.

Noon acknowledges that every land-use decision or resource use involves a trade-off and every short-term benefit entails some immediate or
long-term cost. "Who better than ecologists to assess the environmental costs

of our decisions? I believe we have a clear role, and responsibility, to assess the costs, communicate the trade-offs, and initiate the collaborations," he says.

With the passage of time, both Barry Noon and Ben Santer have experienced vindication and acknowledgment that their science was right even as their struggles continue. "Anger resulting from the continued distortion or suppression of science is still a strong motivator for me, even if it is not very noble," Noon says. Noon is now devoted to training the next generation of young scientists as researchers and as communicators. Santer now devotes 30 percent of his time to public education about climate change.

When you experience backlash, it's important to tune into your family, friends, and trusted colleagues who will remind you that your results and inferences are well founded. Conflict further distorts your perception of how people see you and your work—you'll hear much less from the silent majority of people who support you if you don't seek out their opinions. It's easy to become obsessed with the minority voices of critics while overlooking all the praise and votes of confidence.

The Bottom Line

While dealing with backlash is no fun, many scientists agree that the end result is often worthwhile. Eventually, the truth will prevail. The best defense is doing good science—and communicating it, because "the data do not speak for themselves" (*Nature* 2010).

Chapter 15

LEADING THE WAY:
TEN STEPS TO SUCCESS

Sticking your neck out gets more comfortable with practice.
　　—Larry Crowder

It's great to establish a reputation within academia as the scientist who was clever enough to sort out a complex problem. It can be even more rewarding to see society at large take note because you've communicated its relevance. Or to have neighbors casually tell you about science that *you* had a hand in. When journalists pay attention, the public becomes aware, and policymakers decide it's an issue that can't be ignored. Then something shifts. That's where some scientists realize an even more profound sense of satisfaction: you have literally changed the world.

When you embrace public communication you are assuming leadership. With few exceptions, those who lead the pack are excellent communicators. Leaders know what they want to say, when, how, and to whom. And whether or not it's apparent, they probably work hard at it. In naming Apple Inc. co-founder Steve Jobs "CEO of the Decade," *Fortune* magazine pointed to his attentiveness to communications. "A key Jobs business tool is his mastery of the message. He rehearses over and over every line he and others utter in public about Apple" (Lashinsky 2009).

Good leaders are always in demand—both within and beyond the scientific arena. Honing your communication skills will open new doors for you. Whether you decide to engage with the media, get involved with policymakers, or become a leader within an institution, becoming a better communicator can give you power to advance your goals, help the next generation, or even shift the culture of academia to towards recognizing and rewarding outreach.

If you choose, you can become an agent of change—a scientist who leads the scientific community and society in new directions by your work, your vision, and your ability to communicate it. You can broaden the perspectives of colleagues, and your impact can extend beyond to the public, policymakers, and other constituencies. How you engage is a very personal decision. But of course your foundation is your sound science.

These nine steps to success draw on all the previous chapters to help you apply your communications skills to making a difference. Think of them as check-in points. You may have mastered some of these elements of being an agent of change, while others may push you into the discomfort zone—at least initially. We hope they will prove helpful as you move forward in whichever way you decide is best for you.

1. Resolve to Speak up for Your Science

Joanie Kleypas of the National Center for Atmospheric Research was one of the first scientists to begin researching how higher levels of carbon dioxide in the atmosphere are making the oceans more acidic. In the early days of this work, she was sitting at a large table with colleagues, examining some of the first model results that predicted the changes in the ocean's chemistry. Suddenly it dawned on her: increasing acidification could have profound, even fatal, implications for creatures with calcium carbonate shells and all of the animals that feed on them. She quietly excused herself, locked herself in the bathroom, and threw up.

Ever since, she has focused not only on research but working to raise awareness of the potential catastrophic implications of ever-rising CO_2 levels. "For me, the turning point was when I realized the severity of the problem I was working on, how little attention people pay to it, and that no matter how good the information was, science just does not have a red-line phone to policy," Kleypas says. "I grew up in a society where people always said things like, 'Why don't they fix that?'" Despite her apprehensions, she realized that if she didn't try to help fix it, who would?

Like many researchers, Kleypas lives and breathes her science. Even after this turning point, she periodically lapsed in her efforts to communicate as her ingrained training as a scientist struggled against her new resolve. In 1999, NPR interviewed her about a paper published in *Science* that she had cowritten. The study stated that increasing ocean acidification was inevitable and could devastate coral reefs in the future (Kleypas et al. 1999). Kleypas explains:

> I had just seen some of the headlines coming out of Australia after one of my coauthors had interviewed with the press there. The headline read "Great Barrier Reef Will Crumble." I and my coauthors became alarmed, feeling that this was an overstatement—at least given what we knew then. . . . By the time the NPR reporter phoned me, I was very shy about strong statements. He pushed and pushed and even asked me what I thought about a *direct quote* from our paper that "this could have dire consequences for coral reefs." All I could say was "well, maybe. . . ." Needless to say, the interview was not aired.

It wasn't that Kleypas doubted her own results. She buckled to the familiar fear: What would her peers think? She still regrets the hiccup in her conviction. "Oh, man, I would have done so many things differently that day! But mainly, I would have tried to drop my fear that my scientific colleagues might hear this, and I would have been better prepared with what to say to a general audience."

Since then, Kleypas resolved anew to move the issue forward and has put herself out there, to help policymakers—including the U.S. Congress—understand the problem. She and many other scientists have worked hard at it.

Their efforts have paid off. Ocean acidification has since been identified by policymakers as a significant issue, and federal dollars have begun flowing by an act of Congress, the Federal Ocean Acidification Research and Monitoring (FOARAM) Act (Anon. 2009a). "I feel like the work has been influential with respect to getting Congress informed and serious about ocean acidification—although we still have a long way to go," Kleypas says. She continues:

> I send copies of my papers and reports directly to congressmen, but I feel most effective when I stay in contact with the staffers, meeting them

face-to-face, and sending them easy-to-digest information. I have learned that it is well worth the effort to find a way to get their ear.

2. Set a Goal and Use It to Guide Your Commitments

Scientists already lead hectic lives, juggling many things. How is it possible to take on more—to be available when the media calls, or to meet with policy-makers? Pam Matson, who directs the Aldo Leopold Leadership Program, suggests that budding scientific leaders set clear goals to help guide them in the tough choices as to what to take on and what to let go. This is hardly a novel idea, but it is critically important. Your goal can be small or large, but it should be a clear statement about what you want to accomplish.

Matson, who is also dean of Earth sciences at Stanford, set one big goal for herself many years ago: to try to move the world toward sustainability. In the early part of her career, Matson investigated nitrogen and nutrient cycles. She worked at the intersection of terrestrial ecology, soil science, and atmo-spheric science, and studied the processes that led to emissions of greenhouse gases and pollutants from forests and agricultural systems to air and water. One day, she woke up to the realization that not only were there big prob-lems associated with nitrogen that needed to be addressed, but also that she was in a position to do something about it. "I remember the day it suddenly dawned on me that this may be totally fun and exciting from a research per-spective, but it is very scary from an environmental perspective—and I wanted to not just learn about problems but to solve them," Matson says.

Matson did not set out to become a dean. But she took the job so she would have even more influence. "I realized that I could be much more effec-tive in achieving *my* end goal, and harness the skills and the ability of the whole university." Matson assesses everything in terms of sustainability. "The things I get involved in, I get involved in as a leader because I see them as crit-ical steps or necessities to reaching that goal."

When new opportunities present themselves, she revisits her goal:

> I may have to give up some other things. I basically ask myself: are all of these other things still critical to the direction that I think we need to be going? You can't do everything, so you have to keep asking yourself, what

you are going to gain from those new activities with respect to your big goal?

3. Think Solutions, Not Just Problems

Scientists excel at identifying problems and then subjecting theories, results, and conclusions to painstaking scrutiny. But as scientists unveil what sometimes feels like an unending catalogue of pending environmental doom, the rest of the world is primarily interested in solutions. When journalists interview you about a problem, you can be sure they'll expect to complete their story arc wanting to know how to fix them. And when scientists refuse to discuss solutions or offer personal insights, journalists, the public, and policymakers get frustrated. From their perspective, if the experts can't tell us what to do, then who can?

If you recall the message box, solutions are part of the story, especially in environmental science. So prepare appropriate answers. You don't have to be prescriptive, but it's important to at least make the connection to the solution in an even-handed way.

One solid approach is to talk in terms of likely outcomes of various courses of action. Very often the solution is to set limits, or to scale down or slow the problem. Phrase things in "if-then" statements, such as "society needs to decide that if we want X, there are choices we have to make. These would include considering Y and Z." While you don't have to personally champion any given answer, you should be able to point to different options and what they would imply.

Bob Richmond, a marine ecologist at University of Hawaii at Manoa, is careful to separate his science hat from his citizen hat, but he understands the importance of real-world solutions. "An agent of change focuses on outcomes, not just outputs," Richmond says. "He or she is willing to go beyond the collection of data and publication to address the 'so what?' This may involve the willingness to take some lumps from those who don't agree by attending public hearings and if necessary, supporting appropriate court cases."

Richmond worked for many years on the island of Guam, which is a relatively small community. He found that being a guest on talk radio and cable TV opened up channels of trust with members of the local community. By focusing on solutions and action he made the science accessible to important advocates of environmental protection. In one instance, this helped pass leg-islation to establish a series of marine protected areas. In another, it helped block a permit for an incinerator.

"I became increasingly frustrated that science was often nothing more than the garnish on the buffet of information and competing interests laid out before decision makers, whether at the local land-use commission or U.S. congressional level," Richmond says. "I firmly believe that good decisions re-quire accurate and adequate information, and an informed public can lead the leaders either through engagement or the electoral process. If researchers with appropriate expertise are not willing to go the extra mile, there are plenty of mercenaries that will."

Richmond's motto is, "Always be honest; don't go further than your data allow; and don't be scared to add your informed opinion as someone who is more than a scientist—but is also a spouse, a parent, and a member of society who cares about the legacy left."

4. Embrace Criticism

Boris Worm of Dalhousie University has had plenty of experience dealing with backlash after each media splash. He has taken to heart Benjamin Franklin's adage, "Our critics are our friends, they show us our faults."

In fact, listening to his critics has propelled him from one major paper to the next. *Science* published his most recent paper, with twenty coauthors that were in previously warring camps, on his fortieth birthday.

His philosophy is to "meet and greet the folks who criticize you. Don't shun them, don't run away from them, walk toward them." By understanding your critics' motivations, and helping them to understand yours, you stand not only to smooth over conflicts but to forge new collaborations.

Worm approaches backlash like a dispassionate scientist. He has learned that it is important to rise above emotion and keep the discussion on an in-tellectual plane. "Maybe it's a coping mechanism, but when I get criticized, I always take it as criticism of the analysis. Even if it's meant personally I don't take it personally. I think that's a dead end. The top guiding principle I have is stay with the facts," Worm says. "That, to me, is most important. Don't let yourself say something that you can't back up."

Worm also finds it crucial to rely on his network of collaborators, senior colleagues, and trusted confidantes. You can't do it alone. Everyone needs sounding boards and, sometimes, moral support. "When you have collaborators, talk to them and bounce ideas off them or get their opinions . . . When you are in the middle of turmoil, a calm outside voice can really help. That has worked for me," he says.

Worm takes in the views of opposing camps, listens respectfully, and carefully evaluates criticism:

> When you actually meet people and you talk it out, and you are not emotional about it, it gets very reasonable very quickly. You can see where there is agreement and where there is disagreement and have a real discussion. That can be very fruitful and can lead to new collaborations, and new discoveries as I have experienced. I value thoughtful criticism more than anything else. It makes everyone think more deeply and makes us push harder against the limits of the unknown.

5. Remember the Four Ps: Preparation, Practice, Persuasion, and Passion

The best scientist communicators have one thing in common. They take their performance very seriously and carefully prepare—every time. Preparation and practice enable them to make clear expression look like it is the most natural thing in the world.

Ben Halpern is a young scientist who has found himself in the center of many debates. "I seem to be a magnet for controversy," he says. His research lies at the intersection of many different interest groups; marine ecologists, commercial and recreational fishermen, resource managers, and others all have a stake in what his results have to say. Halpern was first thrust into the spotlight in 2000, while still completing his PhD on the global analysis of the impacts of marine reserves. In the early days he exhibited little flair for communicating his science because he tried to convey too much and talked to everyone as if they, too, were scientists. But he soon figured it out.

"In my opinion, one needs to be a good scientist first, and a good communicator second," Halpern says. For him, good communication involves four Ps: preparation, practice, persuasion, and passion. Preparation takes time and requires one to know their audience and tailor the message appropriately.

Practice is essential for any performance to be good. Persuasion flows from the effective communication of a clear, well-thought-out message. And a passionate delivery of your message ties it all together. As Halpern says, "If you aren't engaged and energetic, why should your audience care?"

Halpern imagines tough questions and then scripts the answers ahead of time. His preparation paid off when he presented his and his colleagues' global map of cumulative impacts on the oceans (Halpern et al. 2008) at a press conference at the American Association for the Advancement of Science (AAAS) annual meeting. One journalist sarcastically challenged him at the outset and the others stiffened waiting for his reaction. His answer ready, Halpern handled the attack with equanimity and good humor. You could feel the crowd relax and tilt in his favor. The journalists were impressed by his relaxed, nondefensive style and asked him to talk about his science and his approach to communicating it at the Society of Environmental Journalists conference later that year.

Being prepared makes you less nervous when you are challenged. You can think better on your feet. If you are scrambling to put together what you want to say on the spot, you'll likely wish you said something different after the fact. "I spend a huge amount of time preparing for interviews. I hope the quality of the stories at least in part reflects that effort," says Halpern.

6. Be Relentless

The path to policy change is a long and winding road. Scientists who do policy-relevant science often meet with fierce resistance—particularly if there are business interests involved. But over time, with patience and persistence, you may see the shift from opposition to acceptance.

Martin Krkošek's series of studies on the impacts of parasites transmitted from farmed salmon to wild salmon began with his PhD and spanned more than four years, numerous publications, and ongoing communication with scientists, journalists, the public, stakeholders, and policymakers. At first he was taken by surprise at the vehement reactions to his research even though he was publishing in top peer-reviewed journals including *Science*, the *Proceedings of the Royal Academy of Science*, and the *Proceedings of the National Academy of Science*.

Krkošek and his coauthors were accused of agenda-driven science and worse. He stayed cool. He wrote responses to critiques in scientific journals and fastidiously addressed every scientific criticism to his work on his website, where he also posted lay translations of his papers. 🖱

He responded to queries from the press and met with government officials and managers to present his science. The experience was grueling for a young scientist who would have preferred to just focus on his research. But it yielded results: a change in policy.

In April 2008, the government of British Columbia placed a moratorium on the expansion of salmon farms on the central and north sections of Canada's west coast because of the potential harm to wild salmon stocks. The results came about "not just because of the science but particularly communicating it effectively," Krkošek says. "I had been wondering if all the outreach was worth it as it consumed so much time, ate me up with stress, and dragged me through the mud during the backlash. Now I see that it was." Krkošek believes that the role of scientists is to supply scientific information and build scientific understanding. "Patience is key, because the replication of science and repeated communication is necessary for science to be heard, understood, and ultimately incorporated in the policymaking process," he says.

7. Cultivate Connections

Scientists need to switch from passive to active, not only in their writing but also in building relationships outside of academia. Every journalist and decision-maker has "go-to" people whom they trust. Becoming that trusted source of information is valuable. "Go-to" scientists are accessible, are responsive, and can explain things in terms that anyone can understand.

Scientists are often surprised when journalists tell them that they welcome hearing from them. Journalists prefer to connect directly with scientists rather than going through intermediaries such as university or agency press officers who don't always understand the science. Often journalists feel like they are brushed off, or their calls are not returned in a timely fashion. They are flattered when scientists reach out.

So, rather than just answering a journalist's call and then hanging up and forgetting about it, consider each new interaction or person you meet as an opportunity. Send a brief e-mail with something you think might be helpful as follow up. Think of other people that the journalist might want to talk to and send their contact information.

A good way to attract new connections is by writing op-ed articles. It does more than just give you a chance to express your point of view. It identifies you as an expert with something interesting to say and generates ripples. Journalists and decision makers may contact you, wanting to know more.

Martin Doyle, an associate professor of geography at the University of North Carolina at Chapel Hill, says:

> Op-eds have made me a player in the state. After the Leopold training I published one a month for three straight months. And as a result I've kind of become the go-to guy for anything relating to environmental restoration and water-related aging infrastructure like dams or bridges. I get calls from the governor's chief of staff and from state senators. I was amazed at how quickly it vaulted me into that realm of things.

If you take the opportunity to cultivate connections, the payoff can be rewarding. Doyle knows the importance of reinforcing those good decisions. "I have worked my tail off here with an environmental reporter for the *Raleigh News and Observer*," he says, including inspiring laudatory news articles to reinforce good policy decisions. "When something good happens, I do everything I can to get the these mid-management people and environmental agencies all over the newspaper and then everyone loves them." An important part of expanding your network is to make others look good. They will remember you for it.

8. Expand Your Definition of Success

Give yourself permission to do public outreach and support other colleagues who engage in such efforts. Encourage younger faculty and graduate students to do the same. Some institutions reward time and success in reaching broader audiences or influencing policy. Others do not. It takes leadership from within your academic or professional community for this cultural shift to happen. Developing a community of support may first begin outside your institution, but it is important to try to bring it home.

A few universities now factor in such communication efforts when deciding on promotions for professors and reviews for tenure. Nevertheless, this culture shift is still in its early stages.

Paige Novak, an associate professor of environmental engineering at the University of Minnesota, struggled with her decision to get involved. "As professors we have so much on our 'to do' list," she says. "We are evaluated for how many papers we publish a year, how much money we bring in, how many students we advise. This becomes part of the conscious and unconscious judging dance that we do at conferences, meetings, et cetera with each other."

At a Leopold training Novak realized that outreach is what ends up making a real difference:

> What probably made most of us get into this area was the desire to have a real impact and to save the world to some extent. You may write fewer papers in your career but outreach may end up making a real difference in policy or the public's view of an environmental issue or their behavior. That's what you're really after, that real career impact. When you're down in the trenches having your publications counted every year by your colleagues and everyone is always comparing how busy they are, you lose sight of this.

You can help build support for scientists who do outreach and the next generation of scientists by building it into the evaluation systems within your institutions. Larry Crowder is advancing this approach at Duke University. He says, "I tell my students that they should seek research they find personally interesting and satisfying, that is also recognized as significant by fellow scientists. But work that excites the general public that attracts reporters, that engages policymakers—now that's like triple word score in Scrabble!"

9. Seize Unexpected Opportunities

Timing is everything. Grabbing opportunities when they present themselves, even when the timing is inconvenient, is a key to success. Good leaders share a balance of discipline and flexibility that allows them to recognize key opportunities and to take strategic risks. You can do this on a smaller scale by having that op-ed primed and ready to go for when the moment and circumstances are right. Or, at the larger scale, it may mean assuming a new career role when an opportunity presents itself, even when it means sacrificing other priorities or projects.

In December 2008, President Obama selected Oregon State University marine ecologist Jane Lubchenco to head the National Oceanic and Atmospheric Administration (NOAA). The call from the White House came as a surprise while she was with her family in New Zealand. The opportunity required a fast decision. Lubchenco knew that if she accepted, it would change her life. It would also send ripples through her professional circle, affecting many of her associates and the various communication and research programs in which she was involved.

She decided to take the leap. "It is both thrilling and daunting," Lubchenco wrote in an e-mail to me reflecting upon her decision. "It's also surreal right now. This seems like an unparalleled opportunity for science, for oceans, for climate. It's not clear how much I can accomplish, but I'm willing to try."

In an article in *Science*, on January 16, 2008, about the appointment, reporter Erik Stokstad neatly summed up Lubchenco's approach to leadership: "Gather lots of data, think deeply, figure out what needs to be done, and then be an advocate for action" (Stokstad 2009a). Lubchenco not only advocated for action—when it came down to it, she walked her talk.

10. Set Your Own Compass

What are the goals that *you* think are most important to achieve? How can you best make your science matter? No matter what path you take, you will encounter conflicting and even daunting opinions. It's important to chart your own course as well as the speed and direction of your engagement.

When Andy Rosenberg first moved into policymaking as the deputy director of NMFS, he assumed that all he had to do was insert science into the political process. Then—voila!—policymakers would make the best science-driven decisions. He assumed wrong. He quickly learned that advice comes from all directions and that science advisers "risked being undermined by more dogmatic and vociferous stakeholders during the policymaking process." That was especially true given scientists' tendency to emphasize uncertainty and their unwillingness to speculate. Rosenberg says, "Emphasizing what we don't know often drowns out what we do know." As a senior manager for the federal government he walked a tightrope. He says, "Science led my logic. I would start by asking: What do we know and what does that mean we should do?" In every case he would then have to consider: "What can be done given the forces at play?"

As he negotiated complex and sometimes controversial situations, he returned to his set point. Rosenberg says, "My compass is trying to make sure that what I do has an impact on issues I care about and is not just to hear myself talk or to be higher profile in one world or another."

Rosenberg believes that to remember your motivations and keep your long-range goals in mind are paramount. It's too easy to get caught up in the push and pull of policy battles. "At times, you may suffer wilting criticism, and other times [you may be] lauded with praise," he says. What's important is to check with yourself to make sure your actions are moving you in the right direction. Fine-tune your instincts as well as your intellect.

In recent years, Larry Crowder has embraced public debates relating to his science, engaging with journalists and policymakers and dragging his students along with him. Crowder feels strongly that scientists should "find one or more career and communication mentors who you can observe, seek constructive criticism from, who will not only coach you but push you out on a limb." He reminds scientists that it took real boldness to consider a career in science because you are always subject to critique, from editors, grant reviewers, colleagues, even your students. "Once your confidence builds a bit you may be ready for the public arena, where criticism can be harsh," he says. "But nothing is more exciting than to see an idea you introduced take root. When your good idea becomes someone else's idea, you've won."

Barry Noon's experience with the spotted owl debate has made him more committed than ever to make the world a better place. "Scientists have a responsibility to communicate their scientific findings and, when asked, to discuss their policy implications. I believe that scientists must find their own comfort zone when it comes to the public communication of science and should not be criticized for their decision."

The Bottom Line

The path toward progress is not always an easy one. You may encounter roadblocks or resistance, or stumble along the way. That's okay. Keep at it. You are in good company in this sometimes difficult, but ultimately rewarding effort to make your science matter. Where are you now and where do you want to go with these new skills? Find the right fit for you. Perhaps poet Mary Oliver posed the critical question best: "Tell me, what is it you plan to do with your one wild and precious life?"

References

Anon. 1989. *The Montreal Protocol on Substances that Deplete the Ozone Layer.* January 1. Available online at http://ozone.unep.org/Publications/MP_Handbook/Section _1.1_The_Montreal_Protocol.

Anon. 2005. Environmental economics: Are you being served? *Economist,* April 21. www .economist.com/sciencetechnology/displaystory.cfm?story_id=E1_PRRGRQJ.

Anon. 2008a. Ocean thermostat can save coral. BBC, February 8, sec. Science/Nature. http://news.bbc.co.uk/2/hi/sci/tech/7234730.stm.

Anon. 2008b. Fishing and conservation: A rising tide. *Economist,* September 19. www .economist.com/displayStory.cfm?story_id=12253181&source=login_payBarrier.

Anon. 2008c. Editorial: Economies of scales. *Economist,* September 19. www.economist .com/opinion/displaystory.cfm?story_id=E1_TNNGNTJS.

Anon. 2008d. Online video and audio: Fishing and conservation. *Economist,* September 19. http://audiovideo.economist.com/?fr_story=da296b22ba6795fbaee93d5950fb 027fc8774bf0&rf=bm.

Anon. 2009a. *Omnibus Public Land Management Act of 2009. H.R. 146.* March 30. www .opencongress.org/bill/111-h146/text.

Anon. 2009b. Rainforest: Burning issues. *Economist,* August 6. www.economist.com/ sciencetechnology/displaystory.cfm?story_id=14164491.

Arrington, Michael. 2009. A little perspective (Digg, Twitter, Facebook). *Washington Post,* TechCrunch. November 4. www.washingtonpost.com/wp-dyn/content/article/ 2009/11/05/AR2009110500336.html.

Associated Press. 2009. Protect, point, pay—An Associated Press plan for reclaiming news content online. Associated Press, August 13. www.niemanlab.org/2009/08/heres -the-ap-document-weve-been-writing-about/.

Belden, Nancy, John Russonello, Kate Stewart, and American Viewpoint. 1999. *Review of existing public opinion data on oceans.* The Ocean Project, February. www.theocean project.org/images/doc/review.pdf.

Bernton, Hal. 2006. Will seafood nets be empty? Grim outlook draws skeptics. *Seattle Times*, November 3. http://seattletimes.nwsource.com/html/localnews/2003340 489_seafood03m.html.

Brumfiel, Geoff. 2009. Science journalism: Supplanting the old media? *Nature* 458 (March 18):274–77.

Bucy, Erik P., and Maria Elizabeth Grabe. 2007. Taking television seriously: A sound and image bite analysis of presidential campaign coverage, 1992–2004. *Journal of Communication* 57, no. 4:652–75.

Coleman, Felicia C., Will F. Figueira, Jeffrey S. Ueland, and Larry B. Crowder. 2004. The impact of United States recreational fisheries on marine fish populations. *Science* 305, no. 5692 (September 24):1958–60.

Compete, Inc. 2009. Top sites—Ranked lists | Page views—September 2009. compete .com. September. http://lists.compete.com/top-sites/page-views/?data=all.

Costello, Christopher, Steven D. Gaines, and John Lynham. 2008. Can catch shares prevent fisheries collapse? *Science* 321, no. 5896 (September 19):1678–81.

Darimont, C. T., S. M. Carlson, M. T. Kinnison, P. C. Paquet, T. E. Reimchen, and C. C. Wilmers. 2009. Human predators outpace other agents of trait change in the wild. *Proceedings of the National Academy of Sciences* 106, no. 3 (January 12):952–54.

Davies, S. R. 2008. Constructing communication: Talking to scientists about talking to the public. *Science Communication* 29, no. 4 (3): 413–34.

Dean, Cornelia. 2005. About the oceans, he says firmly, attention must be paid. *New York Times*, April 26, sec. Science/Environment. www.nytimes.com/2005/04/26/ science/earth/26prof.html.

Dizikes, Peter. 2009. Our two cultures. *New York Times*, March 22, sec. Books/Sunday Book Review. www.nytimes.com/2009/03/22/books/review/Dizikes-t.html?page wanted=all.

Drake, Perry. 2009. Google vs. Facebook. Drake Direct Roundtable. October 18. http:// drakedirect.blogspot.com/2009/10/draft-facebook-article.html.

Egan, Kieran. 2004. Igniting dialogue through storytelling. Workshop presented at the Dialogue Maker's Series, October 29, Simon Fraser University. www.sfu.ca/mecs/ wosk+dialogue+centre/.

Eilperin, Juliet. 2005. Climate shift tied to 150,000 fatalities. *Washington Post*, November 17, sec. World. www.washingtonpost.com/wp-dyn/content/article/2005/11/16/ AR2005111602197.html.

Erickson, Gregory M., Oliver W. M. Rauhut, Zhonghe Zhou, Alan H. Turner, Brian D. Inouye, Dongyu Hu, and Mark A. Norell. 2009. Was dinosaurian physiology inherited by birds? Reconciling slow growth in archaeopteryx. *PLoS ONE* 4, no. 10 (October 9):e7390. doi:10.1371/journal.pone.0007390.

Falk, John Howard, Elizabeth Donovan, and Rosalie Woods. 2001. *Free-choice science education*. London: Royal Society of Medicine Press.

Finneran, Kevin. 2009. A better future through science citizenship. Panel discussion presented at the Two Cultures in the 21st Century conference, May 9, New York Academy of Sciences. www.nyas.org/Events/Detail.aspx?cid=e2409d6d-7674 -4a68-86a3-91a69269db48.

Gamson, William A., and Andre Modigliani. 1989. Media discourse and public opinion on nuclear power: A constructionist approach. *American Journal of Sociology* 95, no. 1 (July 1):1.

Goldston, David. 2008. Getting it across. *Nature*, July 2. www.nature.com/news/2008/080702/full/454016a.html.

Hallin, D. C. 1992. Sound bit news—Television coverage of elections, 1968–1988. *Journal of Communication* 42, no. 2 (Spring):5–24.

Halpern, Benjamin S., Shaun Walbridge, Kimberly A. Selkoe, Carrie V. Kappel, Fiorenza Micheli, Caterina D'Agrosa, John F. Bruno, et al. 2008. A global map of human impact on marine ecosystems. *Science* 319, no. 5865 (February 15):948–52.

Harvell, C. Drew, Charles E. Mitchell, Jessica R. Ward, Sonia Altizer, Andrew P. Dobson, Richard S. Ostfeld, and Michael D. Samuel. 2002. Climate warming and disease risks for terrestrial and marine biota. *Science* 296, no. 5576 (June 21):2158–62.

Heath, Chip, and Dan Heath. 2007. *Made to stick: Why some ideas survive and others die.* New York: Random House.

Hoffrage, Ulrich, Samuel Lindsey, Ralph Hertwig, and Gerd Gigerenzer. 2000. Communicating statistical information. *Science* 290, no. 5500 (December 22):2261–62.

Homer-Dixon, Thomas. 2007. Terror in the weather forecast. *New York Times*, April 24, sec. Opinion. www.nytimes.com/2007/04/24/opinion/24homer-dixon.html.

International Shark Attack File. A comparison of shark attack and bicycle-related fatalities 1990–2005. Florida Museum of Natural History. Annual Risk of Death During One's Lifetime. www.flmnh.ufl.edu/fish/sharks/attacks/relarisklifetime.html.

Joyce, Christopher. 2002. The sensitive alligator. All Things Considered. National Public Radio, May 15. www.npr.org/templates/story/story.php?storyId=1143409.

———. 2007. Scientists study changing seas on Australian island. All Things Considered. National Public Radio, June 25. www.npr.org/templates/story/story.php?storyId=11366580.

Kaufman, Frank. 2007. Mastering the interview. Training Session presented at the Aldo Leopold Leadership Program, June 27, West Cornwall, Connecticut.

Kleypas, Joan A., Robert W. Buddemeier, David Archer, Jean-Pierre Gattuso, Chris Langdon, and Bradley N. Opdyke. 1999. Geochemical consequences of increased atmospheric carbon dioxide on coral reefs. *Science* 284, no. 5411 (April 2):118–20.

Kleypas, Joan A., Gokhan Danabasoglu, and Janice M. Lough. 2008. Potential role of the ocean thermostat in determining regional differences in coral reef bleaching events (February 9). www.agu.org/pubs/crossref/2008/2007GL032257.shtml.

Krkošek, M., J. S. Ford, A. Morton, S. Lele, R. A. Myers, and M. A. Lewis. 2007. Declining wild salmon populations in relation to parasites from farm salmon. *Science* 318, no. 5857:1772.

Krkošek, M., M. A. Lewis, A. Morton, L. N. Frazer, and J. P. Volpe. 2006a. Epizootics of wild fish induced by farm fish. *Proceedings of the National Academy of Sciences* 103, no. 42:15506.

Krkošek, M., M. A. Lewis, and J. P. Volpe. 2005a. Transmission dynamics of parasitic sea lice from farm to wild salmon. *Proceedings of the Royal Society B* 272, no. 1564:689.

Krkošek, Martin, Mark Lewis, John Volpe, and Alexandra Morton. 2006. Fish farms and sea lice infestations of wild juvenile salmon in the Broughton Archipelago-A rebuttal to Brooks (2005). *Reviews in Fisheries Science* 14, no. 1–2 (January):1–11.

Lashinsky, Adam. 2009. The decade of Steve: How Apple's imperious, brilliant CEO transformed American business. *Fortune*, November 5. http://money.cnn.com/2009/11/04/technology/steve_jobs_ceo_decade.fortune/index2.htm.

Leshner, Alan I. 2007. Outreach training needed. *Science* 315, no. 5809 (January 12): 161.

Lubchenco, Jane. 1998. Entering the century of the environment: A new social contract for science. *Science* 279, no. 5350 (January 23):491–97.

Malakoff, David. 2002. Daniel Pauly profile: Going to the edge to protect the sea. *Science* 296, no. 5567 (April 19):458–61.

McDonald, Bob. 2009. How the bees' knees get a grip. *Quirks & Quarks*. Canadian Broadcasting Company, May 23. www.cbc.ca/quirks/archives/08-09/qq-2009-05 -23.html#5.

Miller, Steve. 2001. Public understanding of science at the crossroads. *Public Understanding of Science* 10, no. 1 (January 1):115–20.

Molina, Mario J., and F. S. Rowland. 1974. Stratospheric sink for chlorofluoromethanes: Chlorine atom-catalysed destruction of ozone. *Nature* 249, no. 5460 (6):810–812.

Mooney, Chris. 2009. Lies 2.0. *Discover Magazine*. The Intersection. October 16. http://blogs.discovermagazine.com/intersection/2009/10/16/lies-2-0/.

Mott, Frank Luther. 1950. *American journalism: A history of newspapers in the United States through 260 years: 1690 to 1950*. New York: Macmillan.

Myers, Ransom, and Boris Worm. 2003. Rapid worldwide depletion of predatory fish communities. *Nature* 423, no. 6937 (May 15):280–83.

Nature, ed. 2009. Filling the void. *Nature* 458, no. 7236 (March 19):260.

Naylor, Rosamond L., Rebecca J. Goldburg, Jurgenne H. Primavera, Nils Kautsky, Malcolm C. M. Beveridge, Jason Clay, Carl Folke, Jane Lubchenco, Harold Mooney, and Max Troell. 2000. Effect of aquaculture on world fish supplies. *Nature* 405, no. 6790 (June 29):1017–24.

Nielsen Online provides topline U.S. data and overall online video usage figures for April 2009. http://en-us.nielsen.com/main/news/news_releases/2009/may/Nielsen _online_April_2009.

Nisbet, Matthew C., and Chris Mooney. 2007. Thanks for the facts: Now sell them. *Washington Post*, April 15. www.washingtonpost.com/wp-dyn/content/article/2007/04/13/AR2007041302064.html.

Obama, Barack. 2008. The president-elect's weekly address—12/20/08. December 20, Change.gov. www.youtube.com/watch?v=PMlXNrBxM0g&feature=youtube _gdata.

Overbye, Dennis. 2009. Elevating science, elevating democracy. *New York Times*, January 27, sec. Science. www.nytimes.com/2009/01/27/science/27essa.html.

Papper, Bob. 2009. *TV and radio staffing and news profitability survey 2009*. Radio Television Digital News Association, April 19.

Patz, Jonathan A., Diarmid Campbell-Lendrum, Tracey Holloway, and Jonathan A. Foley. 2005. Impact of regional climate change on human health. *Nature* 438, no. 7066 (November 17):310–17.

Pauly, Daniel. 1995. Anecdotes and the shifting baseline syndrome of fisheries. *Trends in Ecology & Evolution* 10, no. 10 (October):430.

———. 2005. An ethic for marine science: Thoughts on receiving the International Cosmos Prize. Acceptance speech presented at the award ceremony, October 18, Osaka, Japan. www.seaaroundus.org/newsletter/Issue32.pdf.

Pauly, Daniel, Villy Christensen, Johanne Dalsgaard, Rainer Froese, and Francisco Torres. 1998. Fishing down marine food webs. *Science* 279, no. 5352 (February 6):860–63.

Pew Project for Excellence in Journalism, and Rick Edmonds. 2009. *Newspapers.* The State of the News Media 2009. Pew Project for Excellence in Journalism, April 23. www.stateofthemedia.org/2009/narrative_newspapers_newsinvestment.php?cat=4&media=4.

Pew Project for Excellence in Journalism, and Online News Association. 2009. *Online journalist survey.* Survey Report. The State of the News Media 2009. Pew Project for Excellence in Journalism, March 29. www.stateofthemedia.org/2009/narrative_survey_intro.php?media=3&cat=0.

Pew Research Center for the People & the Press. 2008. *Internet overtakes newspapers as news outlet: Summary of findings.* December 23. http://people-press.org/report/479/internet-overtakes-newspapers-as-news-source.

Pielke, Roger. 2007. *The honest broker.* Cambridge, UK: Cambridge University Press.

Popper, Karl R. 1959. *The logic of scientific discovery.* London: Hutchinson & Co.

Raloff, Janet. 2008. Real news: An endangered species. *ScienceNews,* December 3. www.sciencenews.org/view/generic/id/39022/title/Science_%2B_the_Public__Real_News_An_Endangered_Species.

Revkin, Andrew C. 2008. Climate experts tussle over details: Public gets whiplash. *New York Times,* July 29, sec. Science/Environment. www.nytimes.com/2008/07/29/science/earth/29clim.html.

———. 2009. Industry ignored its scientists on climate. *New York Times,* April 24, sec. Science/Environment. www.nytimes.com/2009/04/24/science/earth/24deny.html?_r=3&hp.

Romm, Joe. 2009. Anatomy of a debunking: A project of the Center for American Progress Action Fund. ClimateProgress. October 19. http://test.cp.techprogress.org/2009/10/19/anatomy-of-a-debunking-yes-caldeira-says-superfreakonomics-is-damaging-to-me-because-it-is-an-inaccurate-portrayal-of-me-and-filled-with-many-statements-that-are-misleading-statements-a/.

Saba, Jennifer. 2009. Exclusive: Average time spent increases—at least a bit—at many top newspaper Web sites in July. Editor & Publisher (August 21). www.editorandpublisher.com/eandp/news/article_display.jsp?vnu_content_id=1004005546.

Safina, Carl. 2009. Pew Marine Fellows Meeting. October.

Santer, Benjamin. 2008. Revisiting "A Discernible Human Influence." In *Communicating on Climate Change: An Essential Resource for Journalists, Scientists, and Educators,* ed. Bud

Ward and Sunshine Menezes, 22–23. Narragansett, RI: Metcalf Institute for Marine & Environmental Reporting.

Schneider, Stephen. 2003. Balancing sound science and conservation action. *Society for Conservation Biology Newsletter,* May 5. www.conbio.org/Publications/Newsletter/Archives/2003-5-May/10-2_003.cfm.

———. 2008. "Mediarology"—The role of climate scientists in debunking climate change myths. In *Communicating on climate change: An essential resource for journalists, scientists, and educators,* ed. Bud Ward and Sunshine Menezes, pp. 22–23. Narragansett, RI: Metcalf Institute for Marine & Environmental Reporting.

Shipley, David. 2004. And now a word from op-ed. *New York Times,* February 1, sec. Opinion. www.nytimes.com/2004/02/01/opinion/and-now-a-word-from-op-ed.html.

Stokstad, Erik. 2009a. The transition: Can an ecologist push NOAA up the food chain? *Science* 323, no. 5912 (January 16):321.

———. 2009b. Global fisheries: Detente in the fisheries war. *Science* 324, no. 5924 (April 10):170–71.

Surowiecki, James. 2008. News you can lose. *New Yorker,* December 22. www.newyorker.com/talk/financial/2008/12/22/081222ta_talk_surowiecki.

Thorne, Stephen. 1997. Overfishing, not seals, killed cod, buried fisheries report reveals. *Globe and Mail,* August 22.

Weiss, Kenneth R., and Usha Lee McFarling. 2006. Altered oceans. *Los Angeles Times,* August 30. www.latimes.com/news/local/la-oceans-series,0,7783938.special.

Wells, William G. 1996. *Working with Congress: A practical guide for scientists and engineers.* Washington, DC: American Association for the Advancement of Science.

Witze, Alexandra. 2009. Jane Lubchenco: The new head of NOAA talks priorities. *Nature News,* March 24. www.nature.com/news/2009/090324/full/news.2009.182.html.

Worm, Boris, Edward B. Barbier, Nicola Beaumont, J. Emmett Duffy, Carl Folke, Benjamin S. Halpern, Jeremy B. C. Jackson, et al. 2006. Impacts of biodiversity loss on ocean ecosystem services. *Science* 314, no. 5800 (November 3):787–90.

Worm, Boris, Ray Hilborn, Julia K. Baum, Trevor A. Branch, Jeremy S. Collie, Christopher Costello, Michael J. Fogarty, et al. 2009. Rebuilding global fisheries. *Science* 325, no. 5940 (July 31):578–85.

Wyss, Bob. 2009. New journalism-science initiatives alter how news is shaped. *SEJournal* 19, no. 3 (Fall 2009).

Zimmer, Carl. 2009. How to more effectively communicate science issues to the public. Panel discussion presented at the Two Cultures in the 21st Century conference, May 9, New York Academy of Sciences. www.nyas.org/Events/Detail.aspx?cid=e2409 d6d-7674-4a68-86a3-91a69269db48.

Index

Note: page numbers followed by "f" or "t" refer to figures or tables, respectively.

Interviews (*continued*)
do's and don'ts, 139–41
five Cs (concise, conversational, clever, correct, cool), 126–27
follow-up after, 130–32
Hayden on, 131–32
ingredients of a great interview, 123–24
journalists, quick preparation by, 124
journalists' reactions, watching, 127
not being quoted in the story, 133–34
offhand remarks, 129–30
"on and off the record," 134–36
opposing views and quotations, 138–39
phone calls, unexpected, 124–25
review prior to publication, 136–37
sound bites, 128–29
on TV, 150–51
Worm on preparing for, 118
wrapping up, 130
Issue advocate role, 20
Issue leaders in Congress, 186

Jackson, Jeremy, 135, 136f, 204, 211
Jargon
journalists' perceptions of, 39–40
message and, 111–12
op-eds and, 165
storytelling and, 48–49
Jobs, Steve, 219
Job security, 211
Johnson, Lyndon B., 89
Journalism
competition in, 34
deadlines in, 40, 68
Journalists. *See also* Interviews; Outreach
audience of, 106
at conferences, 154–55
culture of scientists vs., 29–34
as dance partners, 119–20
democracy and, 17–18
feedback to, 132, 161
independence of, 131, 137
inviting into lab or field, 160
knowledge of, 37–38
message box and, 109
multimedia and, 67
perceptions of scientists by, 38–40
quick investigations by, 124
reactions of, watching, 127
relationships, building, 156, 158–60, 227–28
scientists' perceptions of, 35–38
specialized, 68
as storytellers, 41
Journals, high-profile, 171

Journals, peer-reviewed, 15
Joyce, Christopher, 39, 42, 43–44, 134–35, 144–47, 148f

Kestenbaum, David, 37–38
Kickers, 156
Kleypas, Joan (Joanie, 190, 200–201, 202–3, 220–22
Knight, Emily, 96, 100, 185
Knowledge
curse of too much, 107–8, 113
of journalists, 37–38
Knowlton, Nancy, 84
Krkošek, Martin130, 226–27

Lab, inviting journalists into, 160
Language
defining your terms, 39, 112
details, 127
jargon, 39–40, 48–49, 111–12, 165
metaphor and analogy, 116, 146
in one-pagers, 191
in op-eds, 164
for radio, 146
word choice, careful, 140
Leadership
overview, 219–20
connections, cultivating, 227–28
embracing criticism, 224–25
four Ps (preparation, practice, persuasion, and passion), 225–26
goal setting, 222–23, 230
patience and persistence, 226–27
resolving to speak up, 220–22
setting your own compass, 230–32
solutions, not just problems, 223–24
success, expanding definition of, 228–29
unexpected opportunities, seizing, 229–30
"Legislative updates," 187
Leopold Leadership training, xi–xiii, 21, 105
Leshner, Alan, 16
Letters to the editor, 160–61
Library of Congress, 91, 186
Limelight, 15
Lindholm, James, 148f
Lobbying vs. educating, 94
Local politics and information, 98
Loder, Natasha, 37–38, 42, 56, 57–58, 126, 204–5
Lodge, David, 94–95, 97–98
Logic, 115
Long-term issues vs. daily events, 50–51
Lubchenco, Jane, xi–xii, 3, 4, 71, 230
Lynham, John, 179